Praise for *Mobile Applications: Architecture, Design, and Development*

"This book walks readers through the process of creating mobile applications, and identifies the main technologies used in the realm of mobile application development. It also lists many of the 'gotchas' readers would encounter migrating or developing applications. I like the overall approach—*Mobile Applications* covers the topics often overlooked by many moving into the mobile application realm."

—Scott Troxell,
Technology Consultant

"The key distinguishing factor for this book is the architecture approach. All of the other available books on mobile applications are code development focused, rather than architecture focused. I believe there is a strong need in the development community for this type of book."

—John Featherly,
Senior Solution Architect

"This is a useful book for software architects, engineers, and project managers attempting to develop applications for mobile devices. The text is interesting and conversational. I recommend it."

—Rick Kingslan,
Senior Systems Engineer

"In *Mobile Applications*, all pertinent aspects of the topic have been covered. It flows well and is useful for high-level managers, and, given the inclusion of code, also for engineers."

—Randy Santossio,
Principal Software Engineer and Adjunct Professor

Mobile Applications:

Architecture, Design, and Development

Valentino Lee

Heather Schneider

Robbie Schell

About Prentice Hall Professional Technical Reference

With origins reaching back to the industry's first computer science publishing program in the 1960s, and formally launched as its own imprint in 1986, Prentice Hall Professional Technical Reference (PH PTR) has developed into the leading provider of technical books in the world today. Our editors now publish over 200 books annually, authored by leaders in the fields of computing, engineering, and business.

Our roots are firmly planted in the soil that gave rise to the technical revolution. Our bookshelf contains many of the industry's computing and engineering classics: Kernighan and Ritchie's *C Programming Language*, Nemeth's *UNIX System Administration Handbook*, Horstmann's *Core Java*, and Johnson's *High-Speed Digital Design*.

PH PTR acknowledges its auspicious beginnings while it looks to the future for inspiration. We continue to evolve and break new ground in publishing by providing today's professionals with tomorrow's solutions.

PRENTICE HALL PTR

Hewlett-Packard® Professional Books

HP-UX

Cooper/Moore	HP-UX 11i Internals
Fernandez	Configuring CDE
Madell	Disk and File Management Tasks on HP-UX
Olker	Optimizing NFS Performance
Poniatowski	HP-UX 11i Virtual Partitions
Poniatowski	HP-UX 11i System Administration Handbook and Toolkit, Second Edition
Poniatowski	The HP-UX 11.x System Administration Handbook and Toolkit
Poniatowski	HP-UX 11.x System Administration "How To" Book
Poniatowski	HP-UX 10.x System Administration "How To" Book
Poniatowski	HP-UX System Administration Handbook and Toolkit
Poniatowski	Learning the HP-UX Operating System
Rehman	HP-UX CSA: Official Study Guide and Desk Reference
Sauers/Ruemmler/Weygant	HP-UX 11i Tuning and Performance
Weygant	Clusters for High Availability, Second Edition
Wong	HP-UX 11i Security

UNIX, Linux, Windows, and MPE I/X

Mosberger/Eranian	IA-64 Linux Kernel
Poniatowski	UNIX User's Handbook, Second Edition
Stone/Symons	UNIX Fault Management

Computer Architecture

Evans/Trimper	Itanium Architecture for Programmers
Kane	PA-RISC 2.0 Architecture
Markstein	IA-64 and Elementary Functions

Networking/Communications

Blommers	Architecting Enterprise Solutions with UNIX Networking
Blommers	OpenView Network Node Manager
Blommers	Practical Planning for Network Growth
Brans	Mobilize Your Enterprise
Cook	Building Enterprise Information Architecture
Lee/Schneider/Schell	Mobile Applications: Architecture, Design, and Development
Lucke	Designing and Implementing Computer Workgroups
Lund	Integrating UNIX and PC Network Operating Systems

Security

Bruce	Security in Distributed Computing
Mao	Modern Cryptography: Theory and Practice
Pearson et al.	Trusted Computing Platforms
Pipkin	Halting the Hacker, Second Edition
Pipkin	Information Security

Mobile Applications:
Architecture, Design, and Development

Valentino Lee
Heather Schneider
Robbie Schell

www.hp.com/hpbooks

PEARSON EDUCATION
Prentice Hall Professional Technical Reference
Upper Saddle River, NJ 07458
www.PHPTR.com

Library of Congress Cataloging-in-Publication Data

A CIP catalog record of this book can be obtained from the Library of Congress

Executive Editor: *Jill Harry*
Cover Design: *Talar Boorujy*
Cover Design Director: *Jerry Votta*
Editorial Assistant: *Brenda Mulligan*
Manufacturing Manager: *Alexis Heydt-Long*
Manufacturing Buyer: *Maura Zaldivar*
Marketing Manager: *Dan DePasquale*
Composition: *Argosy*

Published by Pearson Education
Publishing as Prentice Hall Professional Technical Reference
Upper Saddle River, New Jersey 07458

Prentice Hall PTR offers excellent discounts on this book when ordered in quantity for bulk purchases or special sales. For more information, please contact: U.S. Corporate and Government Sales, 1-800-382-3419, corpsales@pearsontechgroup.com. For sales outside of the U.S., please contact: International Sales, 1-317-581-3793, international@pearsontechgroup.com.

Printed in the United States of America
First Printing

ISBN 0-13-117263-8

Pearson Education LTD.
Pearson Education Australia Pty, Limited
Pearson Education South Asia Pte. Ltd.
Pearson Education Asia Ltd.
Pearson Education Canada, Ltd.
Pearson Educación de Mexico, S.A. de C.V.
Pearson Education—Japan
Pearson Malaysia, S.D.N B.H.D.

Contents

Foreword

Nick Grattan
Microsoft Regional Director
Author of *Pocket PC, Handheld PC Developer's Guide with Microsoft Embedded Visual Basic*
Hook Norton, Oxfordshire, England.

Over the last few years, the development tools and techniques used for both mobile and desktop/server applications have been converging. This trend started with the Microsoft Foundation Classes and Active Template Libraries for C++ and eMbedded Visual Basic. It has since culminated with the Microsoft .NET Compact Framework and Mobile Web Controls for C# and VB.NET being fully integrated with Microsoft Visual Studio .NET.

Coinciding with this trend has been organizations' desire to include a mobile element into existing and new applications, as the benefits of providing such solutions continue to be recognized and demanded by users. Thankfully, hardware, networking, and wireless advances made over the same period of time have supported the advances in software development technology and have now helped make this desire a reality.

Armed with these new tools, developers have been quick to apply existing development skills learned while building Windows Forms and Web Form applications to the development of mobile applications. However, in doing so, developers have discovered that mobile application development is not just a simple extension of desktop and server application development. There are many architectural, design, and implementation decisions to be made that are specific to the mobile arena and making these decisions requires new skills.

And that's why the book you're now reading is so important. Based on their extensive real-world experience, Lee, Schneider, and Schell, have written a book that explores all facets of mobile application development. Topics discussed range from architecture to design, coding, testing, and project management. The specific considerations of mobile application development are explained, and readers are shown how to migrate their existing skills to this new and fascinating area. Crucial design decisions, such as whether to use a Windows Forms intelligent client or a Web Form thin client, are clearly described and then demonstrated using excellent, relevant case studies.

All members of a mobile development team will benefit from reading this book, either in its entirety or simply by delving into those sections relevant to their function or role.

The future of mobile computing is becoming even more exciting. Mobile devices are continually growing more capable, especially with the advent of cleverly integrated phone capabilities. With better and better wireless networks capable of transferring media in real time, an entirely new breed of applications is now possible. Riding this new wave may be extremely profitable for organizations positioned to take advantage of it. It may be even more profitable for individuals prepared to learn and use these new skill sets.

Preface

The miniaturization of PCs in the form of PDAs, Tablet PCs, and other mobile devices has provided new ways for users to interact with computers. It has also led to a need to deploy and redeploy many existing desktop applications for mobile users.

Such deployment, however, is often fraught with problems, including poor business justifications, expensive development costs, technical and ergonomic issues with the mobile devices, imperfect connectivity, complex application development, and release management problems. Often, today's mobile solutions are neither truly comprehensive nor are their implementations always successful.

In part, this is due to the incomplete presentation of the knowledge needed to develop and deploy mobile applications. Much of the existing literature tends to explain how to develop mobile applications technically, without describing the other factors that influence mobile application development and integration.

As a result, there is a gap in knowledge between technical implementation and more advanced principles of mobile application development, such as business concerns, architecture, the need to integrate mobile applications with existing web and legacy applications, and the management of such efforts.

In this book, we attempt to address the advanced principles of mobile application development and integration while also illustrating the specific details of mobile technologies with several examples. In doing so, we hope that you will be able to use this book as your definitive architectural design reference for mobile application development.

INTENDED AUDIENCE

This book is best suited to readers who have some knowledge of general application development and a programming language, such as Microsoft Visual Basic .NET or C#. Project managers, technical leaders, solution architects, intermediate to advanced programmers, and computer science professors and students may also find this book useful.

ACKNOWLEDGMENTS

Writing this book has been a wonderful experience for us but it has also affected our families, friends, and colleagues. We thank them all for their patience and understanding.

Our editor, Jill Harry, has been a wonderful guide throughout the book-writing journey and we thank her and the Prentice Hall PTR team for their time and contributions. We also thank the technical reviewers and the Argosy team for their diligent reviews of the book. Finally we thank our colleagues at Hewlett-Packard and Microsoft for their input and advice on a variety of topics and technologies.

DOWNLOAD SITE

The code developed and referenced in this book may be obtained through a download from *http://authors.phptr.com/lee/.*

CHAPTER 1

Introduction

No man is an island, entire of itself;
every man is a piece of the continent, a part of the main...
—John Donne

This chapter introduces the subject of mobility to the reader and describes what it means to be mobile. It goes on to describe some of the broad considerations that must be taken into account during mobile application design and development.

1.1 WHAT IS MOBILITY?

Mobility can be defined as the capability of being able to move or be moved easily. In the context of mobile computing, mobility pertains to people's use of portable and functionally powerful mobile devices that offer the ability to perform a set of application functions un-tethered, while also being able to connect to, obtain data from, and provide data to other users, applications, and systems.

In order to do so, a mobile device must have certain characteristics. For example, it must be portable and a user must be able to carry it around with relative ease. A mobile device also has to be highly usable and functional, and allow for easy connectivity and communication to other devices (see Figure 1–1).

Each of these characteristics is important in its own right but none can genuinely be considered overriding. From a user's perspective, the mobile device or solution that offers the "best" mobility will generally have some combination of these characteristics.

Figure 1–1 Mobility characteristics

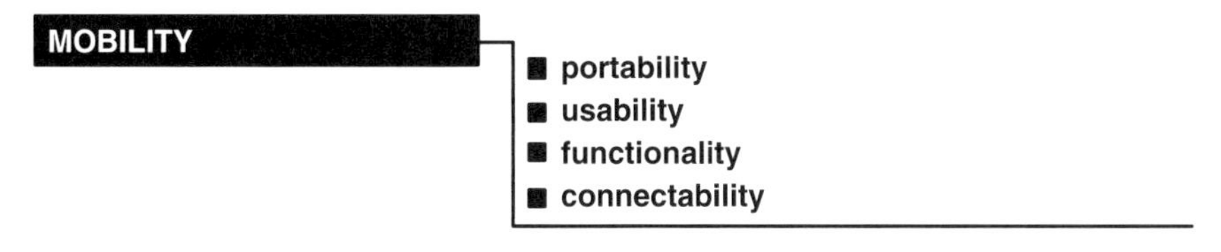

A user may also be flexible about one of the characteristics if there is some corresponding compensatory benefit in another characteristic. For example, a user might allow for some latitude in portability if a mobile device has a great deal of functionality. However, if the device is unusable as a result of its size, the best functionality in the world may not suffice. In the sections that follow, we will discuss each of the mobility characteristics in greater detail.

1.1.1 Portability

Portability is defined as the capability of being easily carried. Today, in order to be considered portable, mobile devices typically have to be easily carried by hand. We say "today" because the definition of portability can change over time. Something that was considered portable in the past may not be considered portable today. For example, early portable PCs developed in the 1980s were as large as small suitcases, while early cellular telephones were the size of walkie-talkies.

Today, devices that are orders of magnitude faster, smaller, and more powerful than those devices of yesteryear can be comfortably carried in the palm of your hand. Tomorrow, continued miniaturization of existing chips and devices may allow for even smaller devices to be made. Advances in nanotechnology may even allow for minute portable devices to be implanted within the body.

That notwithstanding, two of the most important factors affecting the portability of a mobile device include the size and weight of the device and its accessories (see Figure 1–2).

Mobile device size and weight. Size matters! Larger devices are generally harder to carry, while smaller devices are harder to use. It is probably not coincidental that some Laptop PCs, which are arguably the largest devices that can be considered mobile, have shrunk to paper notepad size (approximately 8.5"W × 11"L × 1"H). Anything very much larger would be somewhat cumbersome. At the other end of the spectrum, it is also possible for a mobile device to be too small. For example, certain cellular telephones are so small that they are difficult to use although they are convenient to carry.

The weight of a mobile device is also important. The largest mobile devices today—such as Laptop and Tablet PCs—weigh a few pounds, while smaller mobile devices—such as cellular telephones and pagers—weigh only a few ounces. Today, anything weighing more than a few pounds is unlikely to be viewed as portable.

Figure 1–2 Portability factors

PORTABILITY

- **size of device and accessories**
- **weight of device and accessories**

Mobile device accessories size and weight. Two often overlooked aspects of portability are the size and weight of a mobile device's accessories. These accessories can contribute significantly to the overall portability of a mobile device.

For example, consider the Pocket PC. Advertisements show you how convenient and lightweight it is. It easily fits into the breast pocket of a man's shirt and sits comfortably in your hand.

Now consider the Pocket PC accessories. Typical accessories include a power transformer and cable, a sleeve, a synchronization cable, a cradle, a network communication card, and a carrying case. You might also need to carry an extra-rugged case, a collapsible keyboard, a camera, a barcode reader, a global positioning system (GPS) reader, and extra batteries, depending on what you intend to do with the Pocket PC.

Perhaps in the future even lighter accessories will be made. For now, you should bear in mind that mobile devices require power, connectivity, and accessories to allow them to function in an optimal fashion and that these items also have size and weight to consider.

1.1.2 Usability

A mobile device must be usable by different types of people in many different environments. The usability of a device depends on several factors, including the user, environment, and device characteristics (see Figure 1–3). These are described in more detail below.

User characteristics. A user's interaction with a mobile device depends, to an extent, on his/her personal characteristics. Some of the more common user characteristics include:

> **Size and strength.** A user's size and strength has an effect on his/her interaction with the mobile device. For example, a Laptop PC that is easily carried by an adult may not be usable by a child.
>
> **Flexibility and dexterity.** A user's flexibility and dexterity has an effect on the usability of the mobile device. An adult user may have larger fingers and require a larger keyboard. For

Figure 1–3 Usability factors

USABILITY

- user characteristics
- environment characteristics
- device characteristics

example, a firefighter wearing a thick pair of gloves and trying to retrieve floor plan information from a Personal Digital Assistant (PDA) might find the mobile device unusable.

Knowledge and ability. It is hard to guard against a user's lack of knowledge or ability. Generally, the most usable devices are the simplest and most intuitive to use. If the device is too difficult to master, the user will not find it usable.

Environment characteristics. A user's environment affects the choice of device. Some of the considerations brought about by a user's environment include:

Normal working conditions. A mobile device must work under the user's normal working conditions. For example, if a user has to work standing up for extended periods of time, a PDA or Tablet PC may be more suitable than a Laptop PC since they don't need to be set down.

Extreme conditions. A mobile device may also have to work under extreme conditions (e.g., hot, cold, wet, dry). For example, an emergency services worker, such as a firefighter, working in a hazardous environment may need a waterproofed or more rugged mobile device that can handle the environment.

Device characteristics. Mobile devices have different inherent characteristics, which can affect overall usability. Some of these inherent characteristics include:

Startup time. If a user needs to use a mobile device in a time-critical fashion (e.g., emergency services workers), a mobile device that starts up immediately might be very beneficial. For example, Pocket PCs take only a second or so to start up, while Tablet PCs and Laptop PCs may take minutes to start up.

Data integrity. If a user cannot tolerate data loss and requires permanent, built-in data storage, a Tablet PC or Laptop PC may be preferable to a Pocket PC.

User interface (keyboard, stylus, mouse, etc.). A device's intrinsic characteristics may make it impossible to perform certain functions because of the nature of the user interface. For example, it is very difficult to create freehand drawings using a mouse on a Laptop PC. It is much more comfortable and natural to use a pen on a Tablet PC.

Robustness/ruggedness. Mobile devices are generally not highly robust and may break if they are dropped. Mobile devices can be made more rugged to help them withstand the stresses of a user's environment by placing them in a protected thick plastic case. Making a device more robust, however, can adversely affect its portability and usability.

1.1.3 Functionality

Mobile devices serve multiple purposes and have wide and varied types of functionality. The functionality is implemented in the form of a mobile application and mobile devices typically have multiple mobile applications running on them.

In general, mobile applications fall into two categories—those that can operate standalone (i.e., without any contact with another user or system) and those that are dependent and need to connect to another user or system (see Figure 1–4). Some examples include:

Standalone

- Clock
- Games
- Calculator

Dependent

- Calendar
- Email
- Schedule
- Contacts
- Tasks
- News
- GPS

For the purposes of this book, our interest lies mainly with the design and development of custom mobile applications that function in a standalone fashion for a period of time but ultimately depend on other systems for uploading and downloading information.

1.1.4 Connectability

Like man, no mobile device is an island. Mobile devices have neither the power nor the function to operate entirely alone for extended periods of time.

Thus, while many mobile devices have standalone applications (e.g., clocks, games, and calculators) that allow users to operate standalone for periods of time, their primary function is to connect people and/or systems, and transmit and receive information (e.g., calendars and email).

Figure 1–4 Functionality categories

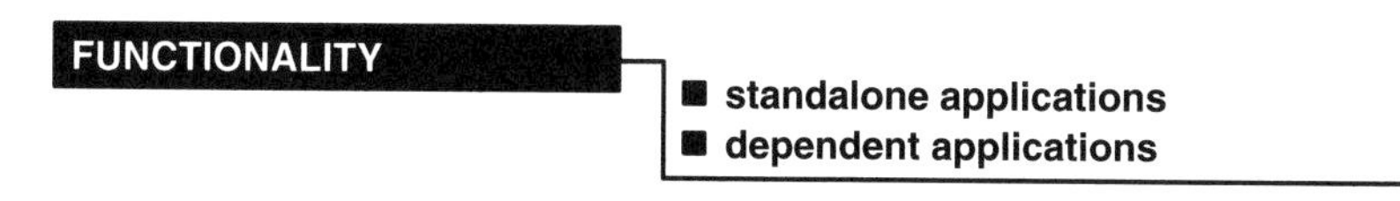

Figure 1–5 Connectability modes

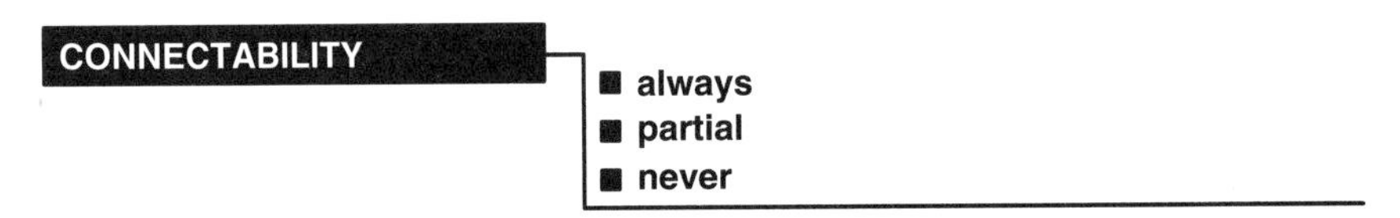

Mobile devices typically operate in one of three modes (see Figure 1–5). A mobile device can always be connected to a back-end system. Alternatively, a mobile device can be intermittently connected to a back-end system. Finally, a mobile device can operate entirely without connection to a back-end system. The first two modes are of critical importance, we will describe them in more detail later.

It is important to differentiate between being *mobile* and being *wireless*. While the two words are commonly used interchangeably, mobility does not necessarily mean having a wireless connection. It is perfectly possible to function in a mobile fashion and be completely un-tethered while gathering information before connecting to a network either wirelessly or in a wired fashion to communicate that information.

1.2 DEVELOPING MOBILE APPLICATIONS

In the following sections, we discuss some of the considerations that must be taken into account when developing mobile applications. We also describe certain fallacies about the mobile application development process.

1.2.1 Umbrella Considerations

It is important to realize that a mobile application is not designed, developed, and deployed in a vacuum. Mobile applications are implemented for business reasons, such as improved productivity, increased accuracy, and other metrics. They also typically need to be integrated with existing applications.

In order to successfully develop a complete mobile application solution, multiple factors must be considered (see Figure 1–6). These considerations, which also form the book's structure, are outlined below.

Mobility. In Chapter 1 (this chapter), we describe what constitutes mobility and a mobile solution.

Business context. In Chapter 2, we address the business context and business rationale behind implementing a mobile solution.

Figure 1–6 Mobile solution

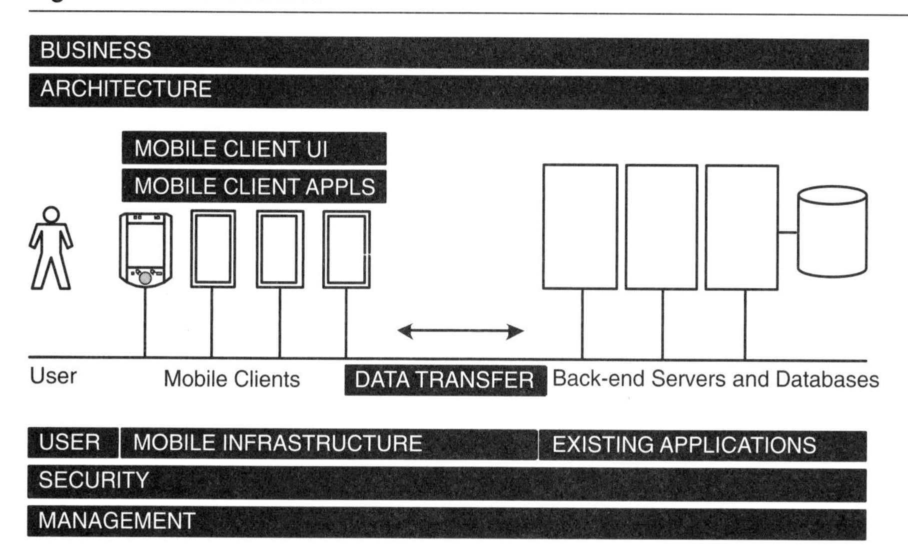

Mobile application architectures. In Chapter 3, we describe several different mobile application architectures and discuss the pros and cons of each for solving different types of problems.

Mobile infrastructure. In Chapter 4, we discuss mobile infrastructure including mobile device types and their components. We will also examine some of the methods that allow mobile devices to connect to a network.

Mobile client user interface. In Chapter 5, we describe the mobile client user interface. This includes the user interface devices, the application content, the user's expectations, and some of the best practices in designing and developing mobile user interfaces.

Mobile client applications. In Chapter 6, we discuss the development of mobile client applications using fat and thin client architectures.

Client-server data transfer. In Chapter 7, we present several different mechanisms for transferring data between the mobile client and the back-end server systems.

Mobilizing existing application architectures. Developing a mobile client application is only part of the whole solution. In Chapter 8, we discuss mobilizing existing web and legacy applications and the considerations that must be taken when doing so.

Security. In Chapter 9, we address security in mobile applications. We have broken this out into a separate chapter because security is pervasive throughout a mobile application. We discuss items such as user authentication, data encryption, communication encryption, and a data self-destruct mechanism.

Mobile application development management. In Chapter 10, we cover mobile application development management. Mobile application development requires project management similar to that needed for major enterprise application and integration projects. We describe some of the best practices for managing mobile application development, including gathering requirements, designing, developing, integrating, testing, and releasing an application. We also outline ongoing maintenance and releases.

Mobile Application Case Studies. In Chapters 11, 12, and 13 we present three complete mobile applications. The source code for these applications is available from this book's companion web site.

In Chapter 11, we discuss mobilizing an existing web site for an art museum. This involves designing and developing a set of thin-client web pages that can be viewed from PDAs and cellular phones that have a continuous network connection.

In Chapter 12, we design and develop a fat-client mobile application for field biologists who enter water sample information into a Pocket PC or Tablet PC while disconnected from the network. When a connection is established, the information is uploaded and synchronized with a back-end web application and database.

In Chapter 13, we develop a web-based tour of a zoo for Pocket PC users who are not connected to a network. This application uses an innovative mechanism to host web pages on the Pocket PC device itself.

1.2.2 Fallacies About Mobile Application Development

We sometimes find that when we present mobile application development proposals and statements of work to our prospective customers there is a sharp intake of breath as they read the effort, time, resources, and price required to mobilize their workforce.

Upon reflection, however—especially when everyone begins to realize that the mobile application requires genuine application development expertise—everyone seems to calm down.

This initial reaction is hard to explain, but it seems to be due to several common fallacies, which can be summarized as follows.

1. **Mobile application development is easier.** People seem to feel that developing applications for mobile devices is easier. In fact, it is probably harder. There are many difficulties that need to be overcome, including ergonomic, connectivity, and smaller display area considerations.
2. **Mobile application development is faster.** There is the notion that developing applications on mobile devices is somehow faster. In actuality, it is probably no faster or slower than any other application development effort. It depends on the complexity of the application being developed and other factors that typically comprise project development.
3. **Mobile application development is cheaper.** Neither the development of mobile applications nor the devices are necessarily inexpensive. The devices themselves are not cheap if

you compare the cost of a Pocket PC or a Tablet PC with a connected desktop computer. By the time you have finished purchasing a Pocket PC and all its accessories, it may be as expensive as a desktop (and may be considerably higher). Part of this cost is simply the intrinsic expense of manufacturing these technologically advanced devices. In addition, you may also need to develop and test a new mobile application on multiple devices, which adds to the overall cost.

1.3 SUMMARY

Mobility pertains to people's use of portable and functionally powerful devices. These devices allow users to perform a set of application functions un-tethered while also being able to connect to, obtain data from, and provide data to other users, applications, and systems.

While people may want true mobility, the technology is not yet there in terms of portability, functionality, usability, or connectivity. Mobile devices that are lightweight, flexible, and useful by one measure are still large and cumbersome by other measures and cannot operate for long periods without the need to obtain power. Extending desktop client, web, and legacy application functionality to mobile devices is still very much in its infancy. We also cannot always guarantee continuous, uninterrupted connectivity.

It is not easy to design, develop, and integrate mobile applications well. As designers and developers, however, this is a fertile area for imaginative and creative solutions.

CHAPTER 2

Business Context

O brave new world that has such people in it.

—William Shakespeare

This chapter addresses the business context behind mobile application development. We initially discuss who is going mobile and what people want to be able to do with their mobile devices. We then discuss the business reasons behind why you might want to mobilize your enterprise.

2.1 WHO IS GOING MOBILE?

The convenience and powerful functionality offered by mobile devices, such as cellular telephones, has resulted in many people already being mobilized. Two of the largest groups of people that use mobile devices are workers and consumers (see Figure 2–1). These are described in more detail below.

2.1.1 Workers

Many enterprises have already initiated serious efforts to mobilize their workforce and many more are making the effort every day. The types of workers that have been identified as primary candidates for mobilization include Road Warriors, Sales and Services personnel, Consultants, Corridor Warriors, and even certain Back-Office workers (see Figure 2–2). Each of these worker types is described in more detail below.

Figure 2–1 Who is going mobile?

WHO?

- **workers**
- **consumers**

Figure 2–2 Worker types

Road warriors. The term "road warrior" refers to employees who travel for much of their work life and who often stay in hotels and use various modes of public transportation. Company managers in many industries might typically fall into this category of worker. While on the road, a manager's business activities might typically include:

- Checking email, schedules, status
- Attending meetings
- Managing staff remotely
- Reading/writing business documents
- Updating email, schedules, status

This type of worker's lifestyle is frequently "on the go." As a result, there are a variety of mobile devices that might suit this type of lifestyle, including cellular telephones, Research In Motion (RIM) devices, PDAs, Tablet PCs, and Laptop PCs.

Sales. There are sales professionals operating within nearly all industries. While selling is a complex and highly specialized business, there are also ordinary activities that sales people will commonly perform, including:

- Preparing for customer meetings
- Traveling to customer sites
- Meeting with customers
- Reading/writing business notes
- Performing follow-up tasks
- Performing administrative tasks

Mobile technology can assist the sales professional in all these areas. For example, preparation for a customer meeting might involve downloading customer account information, product information, or other documentation onto a mobile device. Then, while the sales professional is traveling, a navigational aid (such as GPS) on a mobile device might assist the sales professional in finding the customer's location. During the face-to-face meeting with the customer, a mobile device might help the sales professional show the customer certain new product information or the status of the customer's orders. The sales professional might also be able to enter new orders, update information, or perform administrative activities through the mobile device.

While it is impossible to cover all eventualities, some suitable devices for this type of worker might include cellular telephones, RIM devices, PDAs, and also possibly Tablet PCs and Laptop PCs.

Services. Many industries and professions provide services to customers at one level or another. Our definition for this category of worker is simply that they regularly travel from their own company to provide a service to a customer.

Thus, this category includes "break-fix" type professionals (people who are called out to fix broken computers, cars, etc.), utility meter readers, and emergency services professionals (such as fire, police, and medical workers). While there are numerous types of services, there are also several activities that are common to all services, including:

- Receiving service call
- Obtaining work order information
- Traveling to customer service sites
- Determining what to do
- Performing services or initiating follow-up
- Billing
- Scheduling follow-up tasks
- Performing administrative tasks

Again, mobile technology can assist the services professional in all these areas. For example, the services employee receiving a call may do so over a pager, a RIM device, or a radio. He/she may also download or receive pertinent information about the service call (e.g., malfunctioning part, fire, etc.). The services employee then travels to the service site. Once on site, he/she determines what needs to be done. This might involve referring to documents or information stored on a mobile device. The services professional will then perform the service, fix the problem, or initiate the process that results in the problem getting resolved (e.g., order replacement parts). Once the task is completed, the services professional might fill out billing information, schedule follow-up tasks, and carry out administrative tasks, such as filling out time reports on the mobile device and uploading the information to a back-end system.

Some suitable devices might include cellular telephones, RIM devices, PDAs, and Tablet PCs. Laptop PCs might also be useful but a Tablet PC might be preferred, especially if the job requires a great deal of activity.

Consultants. In this category, we include people such as computer, healthcare, and other specialized consultants. (Consulting is also a type of service, but the usual sequence of events is somewhat different from that described above. The work environment is also typically more stable and the time to provide the service is longer.) Typical activities consultants carry out include:

- Working at customer sites or own office
- Determining what to do
- Performing services or initiating follow-up
- Billing
- Scheduling follow-up tasks
- Performing administrative tasks

Suitable devices might include cellular telephones, RIM devices, PDAs, Tablet PCs, and Laptop PCs, depending on the consultant's mobility.

Corridor warriors. The term "corridor warriors" refers to office workers who have to move around an office a great deal or perhaps move to another office or conference room for meetings on another floor or nearby site. They may frequently have to type or take notes on a device from different locations.

Suitable devices might include PDAs, Tablet PCs, or Laptop PCs.

Back-office. Back-office workers work within a company and normally do not interact face to face with customers. These employees might include system administrators and human resources and warehouse personnel.

Generally, this type of worker's lifestyle is not highly mobile, although there are a few notable exceptions. For example, warehouse personnel might use mobile devices to scan inventory. As a result, these workers might benefit from PDAs or Tablet PCs coupled to a barcode scanner.

2.1.2 Consumers

A large number of consumers can also be expected to use mobile devices and mobile solutions. Two types that we will consider are the young and adult mobile consumers (see Figure 2–3).

Figure 2–3 Consumer types

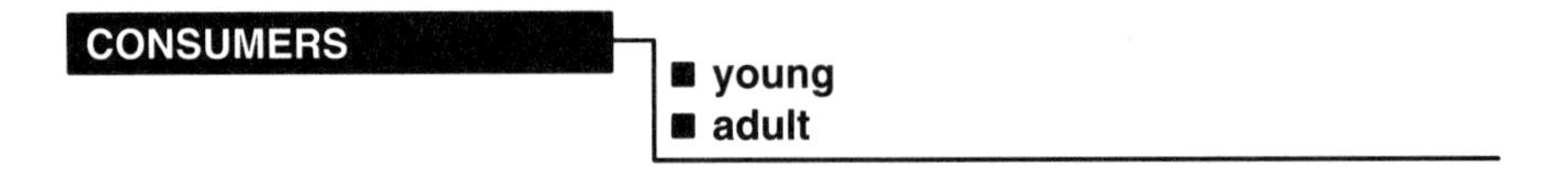

Young and adult consumers tend to use cellular telephones and one or more of a combination of PDAs, Tablet PCs, and Laptop PCs, primarily for home, school, and leisure-oriented activities. The two consumer types are described below.

Young. Young consumers can be expected to utilize mobile devices for communication, entertainment, and education purposes. Communication includes ordinary cellular telephone usage along with text messaging. Entertainment includes the use of mobile devices for games, music, and movies. Education includes reading books and other information downloaded from the Internet and other information provider sources.

Adult. Adult consumers include off-duty professionals and technologists, gamers, shoppers, and entertainment buffs. These people also utilize mobile devices for communication, entertainment, and education purposes. Here, however, communication is generally ordinary cellular telephone usage with much less text messaging. Entertainment includes the use of mobile devices for games, music, and movies. Education includes reading books, news, and other information downloaded from the Internet and other information provider sources.

2.2 WHAT DO PEOPLE WANT TO DO?

There seems to be no end to the ingenious uses for mobile solutions. The primary uses, however, seem to be as a means of communication and for work, entertainment, education, and location purposes (see Figure 2–4).

In the sections below, we describe some of the uses to which people put mobile devices and highlight the more innovative of these uses with examples from several industries.

2.2.1 Communication

Probably the most fundamental thing a mobile user wants to do is communicate. Communication, in this context, refers to the ability to exchange voice, audio, text, and image information (see Figure 2–5). These types of communication are described in more detail below.

Figure 2–4 What do people want to do with mobile solutions?

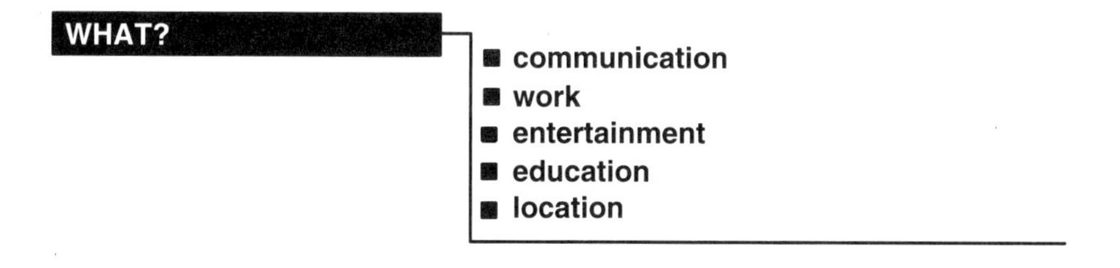

Figure 2–5 Communication types

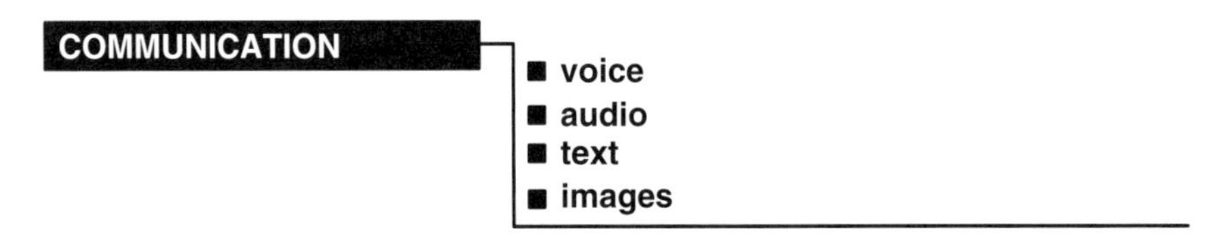

Voice. Users communicate with each other through voice using mobile devices such as cellular telephones (no real surprise here).

Audio. Users also use mobile devices to play and listen to music.

Text. Users can receive and send text data to one another and to other systems through mobile devices.

Images. Users can also receive and send image data to one another and to other systems. For example, a popular pastime is the use of cellular telephones to transmit photographic images to friends and family.

2.2.2 Work

Workers want their mobile device or mobile application to make their work life easier or better in some fashion. Mobile devices provide several ways to do this, including simplifying the exchange of information and the issuing of instructions (see Figure 2–6). These are discussed in detail below.

Exchange information. The types of information exchanged (i.e., obtained and updated) between mobile clients and an application running on a server is very wide. Below we outline some common types of information that users typically interchange:

- Email and instant messages
- News
- Schedules
- Tasks
- Assignments
- Flight information

Figure 2–6 Work uses

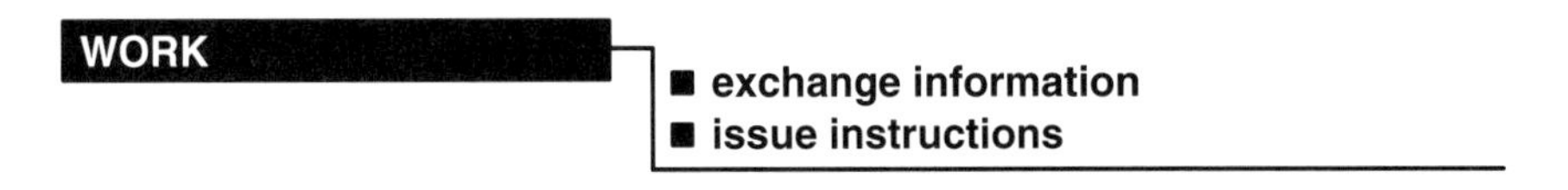

- Geographic information
- Orders and bills
- Package delivery information
- Medical data
- Other work-related information

Mobile workers exchange this information in a variety of ways. For example, many business users use their mobile devices to retrieve email, news, schedules, tasks, assignments, flight and other information from a variety of public locations (such as airports and coffee shops), often through the use of a wireless "hotspot."

In the healthcare industry, mobile healthcare workers and emergency medical services workers have also found many uses for mobile applications. For example, doctors, nurses, and other mobile healthcare workers can update patient records while doing their rounds using Pocket PCs and Tablet PCs. In addition, emergency services workers can use language translation software running on Pocket PC devices to assist them in situations where patients do not speak English.

In law enforcement, mobile applications allow police to access state and federal databases, which contain vehicle registrations, drivers' licenses, and warrant information, over a wireless network using a PDA.

The U.S. government has also been putting mobile devices through their paces with a variety of highly innovative uses. For example, Pennsylvania Department of Environmental Protection field agents use Pocket PCs and ESRI software to enter geographic data on locations where animals infected with West Nile virus have been found. This data can later be analyzed and used to help track and monitor the spread of the disease.

U.S. Armed Forces also use special rugged PDAs and other portable PC devices for a variety of purposes, such as the display of maintenance manuals, determining troop locations, and transmitting firing coordinates.

Finally, one of the most familiar uses is that of package and mail delivery (e.g., Federal Express and UPS). People often sign for a package on a mobile device screen. This data is then uploaded to the courier's back-end systems, allowing the sender to see that the package has been delivered and signed for.

Issue instructions. The ability to issue instructions to a set of business application systems (e.g., when buying or selling a stock) through a mobile device is extremely attractive because it allows workers to take action while they are out of the office. Below we outline some types of instructions mobile users typically issue:

- Buy/sell orders
- Bank transfers
- Other work-related instructions

In the financial industry, many banks and brokerage houses have implemented mobile solutions that allow mobile users to carry out banking and trading functions and also exchange information. For example, account holders at banks such as Citibank can obtain account information and perform certain activities via a cellular telephone.

Similarly, account holders, brokers, and traders with brokerages such as Merrill Lynch can check on the state of the market by downloading news and alerts or checking email on their PDAs. They can also issue trade buy or sell orders through a mobile application.

2.2.3 Entertainment

Users also use mobile devices for entertainment and leisure purposes. The entertainment industry supports different types of mobile computing, including games, gambling, music, and movies (see Figure 2–7). The following describes some of the more prevalent uses in more detail.

Games. People play games on mobile devices for enjoyment and relaxation, and sometimes just to idle away a moment. A plethora of games are available. Java and Binary Runtime Environment for Wireless (BREW) games, in particular, are growing in popularity. For example, see the Yahoo! Games web site.

Gambling. People can obtain information about sporting venues and place bets through the use of PDAs. For example, see the Hong Kong Jockey Club web site.

Music. Downloading and playing music on small portable devices is also a very popular pastime. Some music can be downloaded for free, while other music can be downloaded through a fee-based music subscription service. For example, see the Apple or Napster web sites.

Movies. Movies can also be viewed on certain mobile devices. A pleasant past time for road warriors is conveniently renting DVDs at certain U.S. airports for viewing on a Laptop PC during the flight. After the flight, the DVD can be returned at a local video store.

2.2.4 Education

Users may also utilize mobile devices for educational purposes. This can range from more passive pursuits, such as reading news, to more proactive pursuits, such as interactive learning (see Figure 2–8). Both of these types of educational uses are described below.

Figure 2–7 Entertainment uses

Figure 2–8 Education types

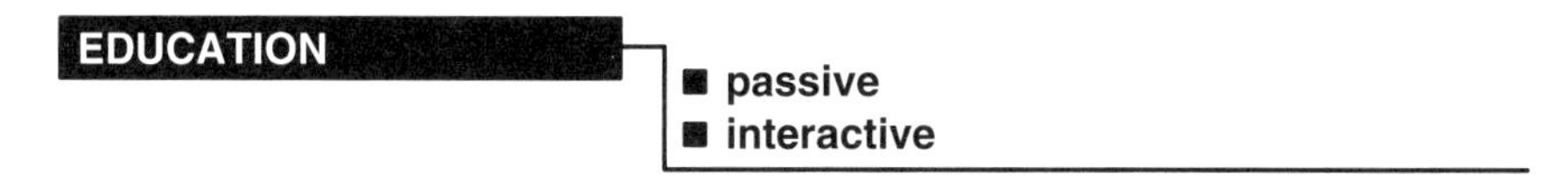

Passive. Mobile devices can provide a less bulky and less wasteful alternative to traditional paper media. Users can read electronic books through a reader or browse the news through a web browser. For example, a subscription to AvantGo allows users to download content from news providers to a desktop computer for synchronization with a PDA. Teachers can display slide presentations to their students rather than using paper handouts. Another benefit, especially to visually impaired users or younger users, is a mobile device's ability to read a story audibly.

Interactive. Users can also utilize their mobile devices for learning in a more proactive and interactive fashion. For example, teachers can use special software from Educational Testing Service to pose questions to students on their mobile devices. Student answers can be immediately sent to the teacher, allowing the teacher to give students immediate feedback and individual help.

2.2.5 Location

Finally, users can utilize their mobile devices for location services. This includes accessing information about a user's current location and also information about a user's destination (see Figure 2–9). Both types of information are described below.

User location information. Mobile devices can provide context-sensitive information about a user's location. This is often accomplished using GPS (if the user is outdoors) or indoor location detectors (if the user is inside a building). For example, electric and gas utilities field workers use custom ESRI software on mobile devices for database access, mapping, global information system (GIS), and GPS integration to help find faults and problems with power and gas utility lines.

Destination information. Mobile devices can also provide users with information about reaching a specific destination. For example, users can download point-to-point driving directions and maps from MapQuest onto a mobile device and follow them. In addition,

Figure 2–9 Location uses

LOCATION

- user location information
- destination information

many luxury automakers (such as Lexus) now offer sophisticated on-board navigation systems that provide dynamic location and direction information to drivers.

2.3 WHY MOBILIZE YOUR ENTERPRISE?

Now that we have a snapshot of who is going mobile and what people are doing, here is a more interesting question: Why would you mobilize your enterprise? Below we discuss some of the pros and cons of going mobile.

2.3.1 Pros

The ability to mobilize and connect people has both subtle and extraordinary consequences in the way we interact with one another and with existing technologies.

Below, we describe several common benefits to mobilizing your enterprise. Some directly affect your employees, while others affect their interaction with your customers and existing business applications, while still others affect your business's finances. Even a small subset of these reasons is a powerful argument for mobilizing your enterprise.

1. **Improve people's lives.** Mobile solutions can help improve people's personal and professional lives. For example, cellular telephones can help working parents communicate with children, child care providers, and schools, ranging from simple logistics to helping better protect children by being in continual contact.
2. **Increase employee flexibility and accessibility.** Providing employees with mobile solutions gives them flexibility from a location and time perspective. By being able to access critical information in the field, employees can move activities—such as the selling process—closer to the customer. In addition, employees may be able to exchange information with the office at a convenient time even when they are away from the office.
3. **Improve employee safety.** Providing employees with up-to-date situation information can improve workers' safety, especially if the employees are working in a hazardous environment.
4. **Improve workflow efficiency and productivity.** Mobilization also helps eliminate redundant activities such as having to re-enter data. For example, if a person ordinarily takes paper notes during meetings, that person may have to re-enter the information on a computer back at his/her desk. This is redundant and error prone since it has to be done twice. Reducing the amount of repetitive and redundant tasks allows employees to be more efficient and productive.
5. **Improve data currency and accuracy.** A mobilized workforce can receive and provide information to existing business systems in a timely fashion. The number of errors can also be reduced during the data gathering and reporting process. Thus, data currency and accuracy are improved.

6. **Improve existing business processes.** A mobilized workforce can be viewed as an additional channel that provides data to and accepts data from existing business systems. In doing so, enterprises may be able to find improvements and eliminate redundancies in existing workflows.
7. **Improve inventory control.** Enterprises can utilize mobile devices to help locate, track, and monitor equipment and other assets. This helps reduce inventory losses and lowers collection costs.
8. **Increase customer satisfaction.** Customer satisfaction may be enhanced once sales and services processes become more efficient and responsive. This, in turn, can lead to increased revenue.

2.3.2 Cons

There are situations, however, where mobilizing your enterprise may not necessarily be appropriate. Below are some of the business, social, privacy/security, and environmental reasons that should be considered.

1. **Business considerations.** Before you mobilize your enterprise, you should have a thorough understanding of the Total Cost of Ownership (TCO), the Total Benefit of Ownership (TBO), and the Return on Investment (ROI) for mobilization. Generally, if the TBO and ROI outweigh the TCO, it would be beneficial to mobilize your enterprise. If the TCO is high or the ROI is low, it is better not to mobilize your enterprise. Factors that contribute to a high TCO include:
   - hardware and software costs
   - communications costs
   - development and deployment costs
   - disruption to existing services
   - operations costs
   - miscellaneous costs (e.g., employee or customer dissatisfaction)
2. **Social considerations.** Being connected all the time also means that people can also reach you all the time, which is not necessarily a good thing. When cellular telephones first became popular, most people thought that they would only use them in emergency situations, such as calling for assistance on the highway because of a car breakdown. However, in the space of a few years, we seem to have gone from occasionally using cellular telephones to becoming completely dependent or even fixated upon them.

   Many people now feel that the excessive use of mobile devices, such as cellular telephones, is rather antisocial. Indeed, cellular telephone usage has been so abused that many public places (such as restaurants) have banned their use.

3. **Privacy and security considerations.** While having information on a mobile device is very convenient, privacy and security considerations need to be addressed when user, company, or customer information is moved to the device. Mobile devices are not highly secure. They are easily lost and the data on them can be easily compromised if care is not taken to secure it.
4. **Environmental considerations.** We often treat mobile devices as though they are disposable. Unfortunately, they really aren't. The fact that they are small does not make their components any less toxic.

   For example, batteries in older cellular telephones use cadmium, while their integrated circuit boards contain a variety of hazardous substances including gold, arsenic, chromium, lead, nickel, silver, and zinc. Many of these substances are linked to cancers and neurological disorders and can have devastating effects when released into the environment.

   Newer battery technology uses lithium ions or nickel metal hydrides, both of which can be recovered and recycled. In addition, methanol or other alcohol-powered fuel cells hold some promise. However, until we can make components less toxic and improve disposal and recycling mechanisms, mobile devices pose a serious problem to the environment.

2.4 SUMMARY

A wide range of people are going mobile for communication, work, entertainment, education, and other reasons. There are many excellent reasons for mobilizing your enterprise. However, you should only proceed to do so if you have a solid understanding of *why* mobilization is beneficial to your enterprise.

CHAPTER 3

Mobile Application Architectures

Where there is no vision, the people perish.
—Proverbs 29:18

In this chapter, we discuss mobile application architectures. We start by describing some of the general concepts and terms behind client-server architectures and follow this by describing clients and servers and the connectivity between them. We then present several interesting architectural patterns and describe why they are useful as general mobile application architecture solutions. Finally, we discuss some of the tenets behind good architectural design and the considerations you need to be aware of when designing mobile applications.

3.1 CLIENT-SERVER

Application architectures are often modeled to highlight or illustrate the overall layout of the software (e.g., application code and platform) and hardware (e.g., client, server, and network devices). While there are many possible combinations of software and hardware, application architectures often fall into a series of recognizable patterns.

Application architectures are commonly modeled in terms of a client-server architecture wherein one or more client devices requests information from a server device. The server typically responds with the requested information (see Figure 3–1).

We can further consider client-server architectures using layers and tiers and the communication between the layers and tiers. Each of these is described in greater detail in the following sections.

3.1.1 Layers

Application code functionality is not necessarily uniform throughout an application. Certain sections of application code are better suited for handling the user interface, while other sections are developed to manage the business logic or communicate with the database or back-end systems.

Figure 3–1 Client-server architecture

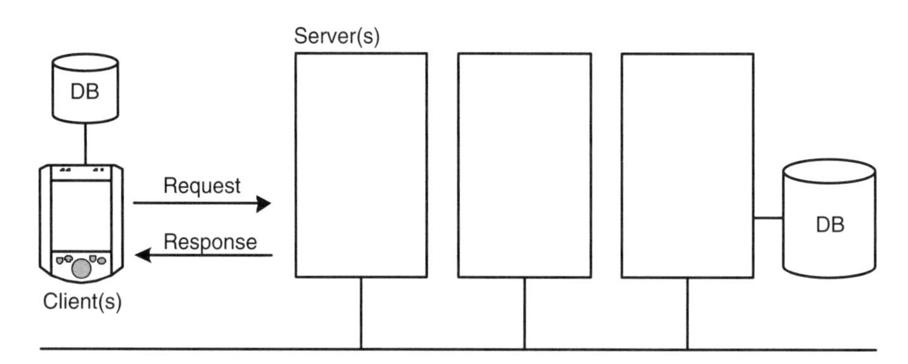

Layering describes the division of labor within the application code on a single machine. Layers are often no more than code modules placed in different folders or directories on the client or server.

With client-side code, there are generally zero to three layers of application code. With server-side code, there are generally one to three layers of application code. This is partly a matter of good software design that helps code re-usability, partly a matter of security, and partly a matter of convenience.

A client with zero code layers essentially has no custom application code. This type of client is commonly referred to as a *thin client* and is possible in client-server architecture if the server holds all the custom application code. A client with one to three layers of application code is commonly referred to as a *fat client*.

A server can also have one to three layers of custom application code. However, you cannot have zero code layers on a server by definition.

The code layer that interacts most closely with the user is often referred to as the Presentation Layer. The second layer is often referred to as the Business Layer, as it typically handles the business logic of the code. The third layer is often referred to as the Data Access Layer. It typically handles communication with the database or data source (see Figure 3–2).

Figure 3–2 Layers

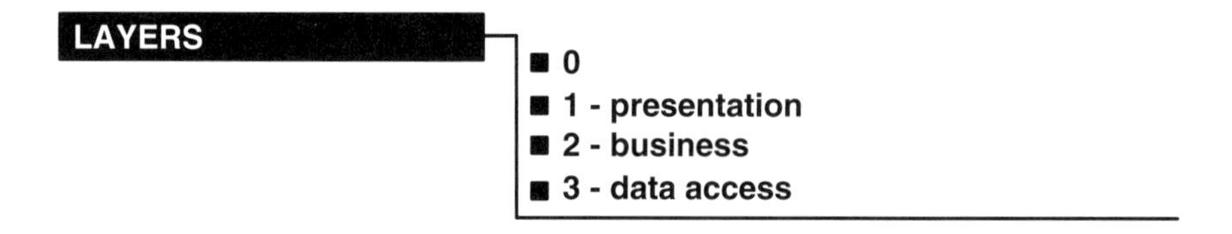

It is possible to have more than three layers on either the client or server but too many layers can become unwieldy and difficult to manage. As a result, this is not frequently encountered.

3.1.2 Tiers

While breaking up application code functionality into layers helps code re-usability, it does not automatically make the architecture scalable. In order to do so, it is important to distribute the code over multiple machines.

Tiers describes the division of labor of application code on multiple machines. Tiering generally involves placing code modules on different machines in a distributed server environment. If the application code is already in layers, this makes tiering a much simpler process.

The code that interacts most closely with the user is often placed in the Presentation Tier. A second tier, which holds the application business logic and data access logic, is often referred to as the Application Tier. The third tier often houses the database or data source itself and is often referred to as the Database Tier (see Figure 3–3). This is an example of a three-tiered architecture.

The servers that make up each tier may differ both in capability and number. For example, in a large-scale distributed web application environment, there may be a large number of inexpensive web servers in the Presentation Tier, a smaller number of application servers in the Application Tier, and two expensive clustered database servers in the Database Tier. The ability to add more servers is often referred to as *horizontal scaling* or *scaling out*. The ability to add more powerful servers is often referred to as *vertical scaling* or *scaling up*. Tiering the application code in such a fashion greatly facilitates the ability to scale applications.

In large-scale distributed web applications, tiers are often bounded by firewalls. For example, a firewall may be placed in front of the Presentation Tier while a second firewall may be placed in front of the Application Tier. The Presentation Tier is thus sandwiched between firewalls in what is termed the Demilitarized Zone (DMZ), while the Application and Database Tier servers are shielded behind the second firewall in what is termed the Intranet Zone. Tiering therefore also facilitates security and allows large enterprises to shield precious internal systems from traffic originating from untrusted zones such as the Internet and DMZ. Without tiering, it becomes very difficult to secure internal systems.

Tiers generally describe server architectures, and we do not typically count client devices as a tier. While it is possible to do so, this is not a usual convention.

Figure 3–3 Tiers

TIERS

- **1 - presentation**
- **2 - application**
- **3 - database**

3.2 CLIENT

There are many mobile device types, including RIM devices, cellular telephones, PDAs, Tablet PCs, and Laptop PCs. These mobile devices can typically operate as thin clients or fat clients, or they can be developed so that they can host web pages (see Figure 3–4). In the following sections, we describe these client types in more detail.

3.2.1 Thin Clients

Thin clients have no custom application code and completely rely on the server for their functionality (see Figure 3–5). Thus, they do not depend as heavily on the mobile device's operating system or the mobile device type as fat clients.

Thin clients typically use widely available web and Wireless Application Protocol (WAP) browsers to display the following types of application content pages:

- Web (e.g., HTML, XML)
- WAP (e.g., WML)

Figure 3–4 Client types

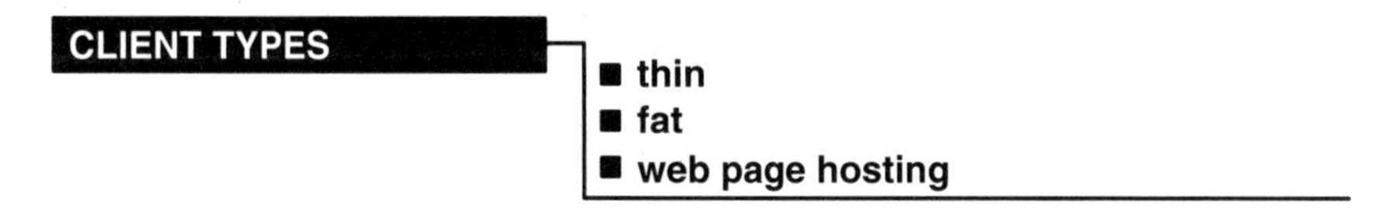

Figure 3–5 Thin client–Zero layers

For example, if web pages are to be displayed, a Pocket PC can display them through Microsoft Pocket Internet Explorer, while a Tablet PC and Laptop PC can also display them through Microsoft Internet Explorer or Netscape Navigator. Similarly, a WAP browser on a cellular telephone can display WML pages.

Thin clients have several advantages over fat clients. For example, they are much easier to maintain and support since there is no application code or data on them. As a result, there is no need to consider application code release and distribution mechanisms to the client.

The problem with thin clients, however, is that they essentially must be in constant communication with the server, since that is their source for updating and obtaining data. If communications are not reliable, you may need to consider standalone fat client applications instead.

3.2.2 Fat Clients

Fat clients typically have one to three layers of application code on them and can operate independently from a server for some period of time.

Typically, fat clients are most useful in situations where communication between a client and server cannot be guaranteed. For example, a fat client application may be able to accept user input and store data in a local database until connectivity with the server is re-established and the data can be moved to the server. This allows a user to continue working even if he/she is out of contact with the server.

However, fat clients depend heavily on the operating system and mobile device type and the code can be difficult to release and distribute. You may also have to support multiple code versions over multiple devices.

Fat clients can be implemented using one, two, or three layers of application code. However, if you only use one layer it is extremely difficult to isolate the individual areas of functionality and reuse and distribute the code over multiple device types. Thus, it is generally better to use two or, preferably, three layers so that you can reuse as much of the application code as possible (see Figure 3–6, Figure 3–7, and Figure 3–8).

3.2.3 Web Page Hosting

It is also possible to display and service web pages on the mobile device even when the mobile client is only periodically connected to the network and back-end systems. In order to do so, we need the equivalent of a "mini" web server on the mobile device.

Figure 3–6 Fat client–One layer

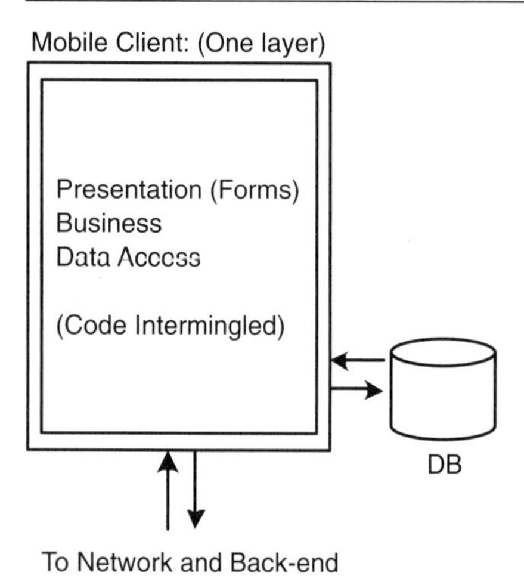

Figure 3–7 Fat client–Two layers

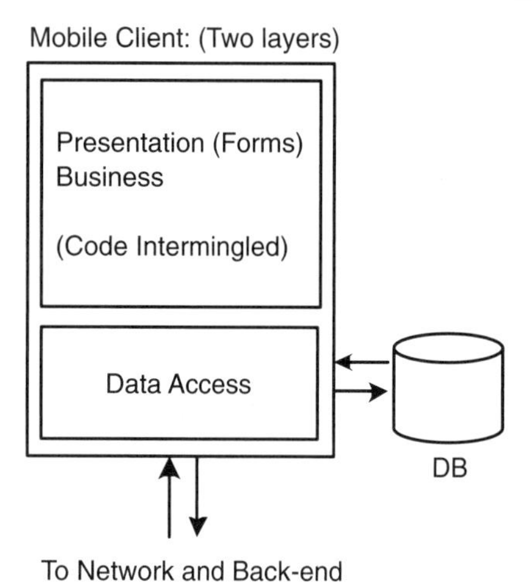

Figure 3–8 Fat client–Three layers

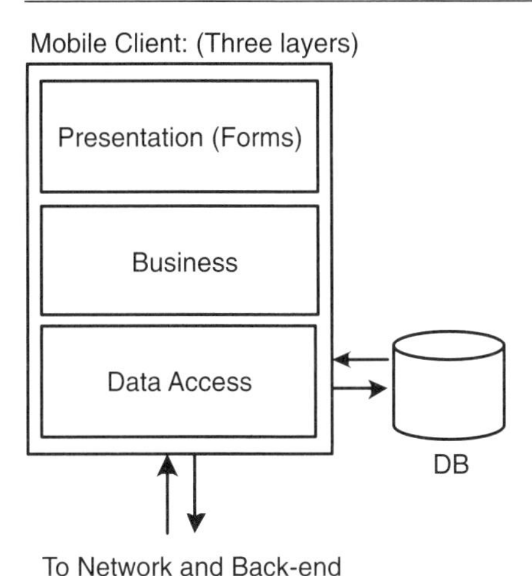

Microsoft has released an HTTP server that runs on a Pocket PC for just such a purpose. Data entered by a user on a web page is serviced by the HTTP server and stored in a local database until it can be uploaded to a server when connectivity has been restored.

Clients that utilize web page hosting can also have one to three layers (see Figure 3–9, Figure 3–10, and Figure 3–11). The main difference between web page hosting and the

Figure 3–9 Web page hosting–One layer

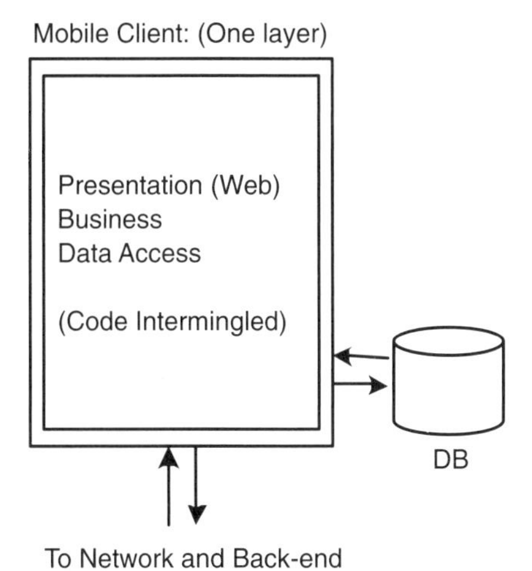

Figure 3–10 Web page hosting–Two layers

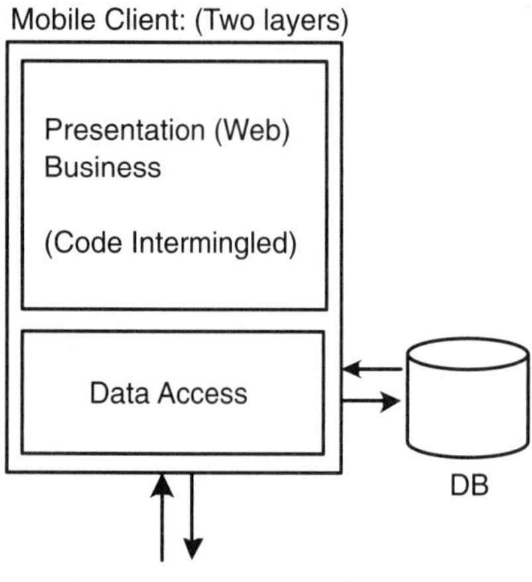

Figure 3–11 Web page hosting–Three layers

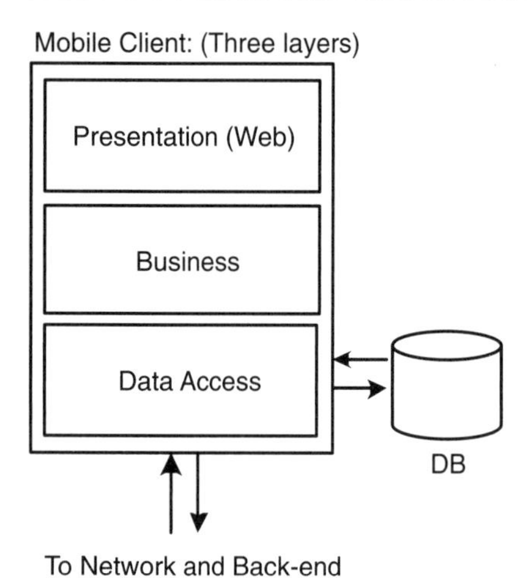

Windows Forms fat client is that the Presentation Layer displays and utilizes web pages instead of Windows Forms.

3.3 SERVER

Server architectures are commonly composed of one to three code layers implemented in one to three tiers. While the temptation is to always build three-tier architectures, there are pros and cons to doing so. For example, large-scale three-tier architectures can be expensive to implement. If the actual application is for a limited number of people, it can be overkill.

In the following sections, we will discuss one-tier, two-tier, and three-tier architectures and some of the pros and cons of implementing them.

3.3.1 One-Tier Architecture

A one-tier architecture can be developed so that three code layers exist on a single server (see Figure 3–12). There are several pros and cons for doing so, as follows:

Pros

- Very convenient
- Quick to develop and deploy

Cons

- Less scalable
- Hard to secure

It is extremely convenient to be able to develop code on a single machine. On the other hand, it is extremely difficult to scale the application. For an internet application, it is also hard to shield the server using firewalls and security zones since you almost certainly have to place the server in the DMZ, which may result in exposing your database to an unacceptably high security risk.

Figure 3–12 One-tier architecture

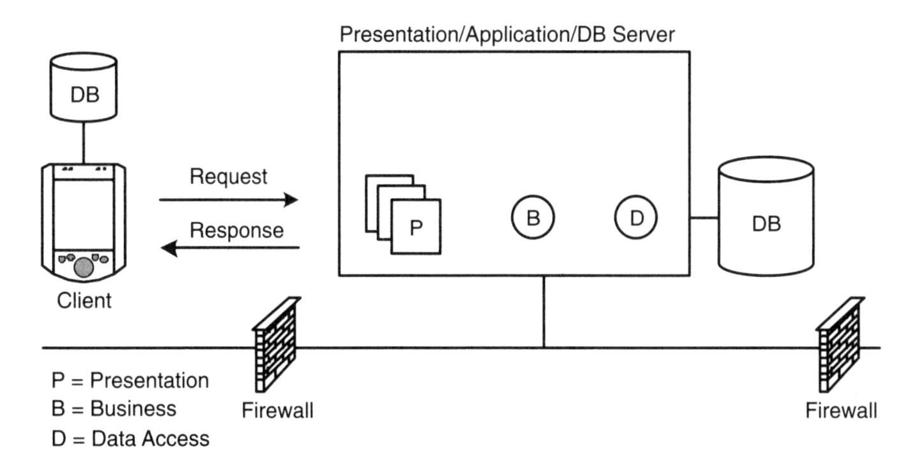

3.3.2 Two-Tier Architecture

A two-tier architecture can be developed so that a database server is split off from the presentation/application server (see Figure 3–13). There are several pros and cons for doing so, as follows:

Pros

- Convenient
- Allows database server specialization

Cons

- Less scalable
- Hard to secure
- More expensive

Splitting off the database server allows it to become a more specialized server. However, it is still extremely difficult to scale the application. It is also still difficult to shield the servers with firewalls and security zones, although this is markedly better than in a one-tier architecture. Nonetheless, you may still expose your application to unacceptably high security risks.

Figure 3–13 Two-tier architecture

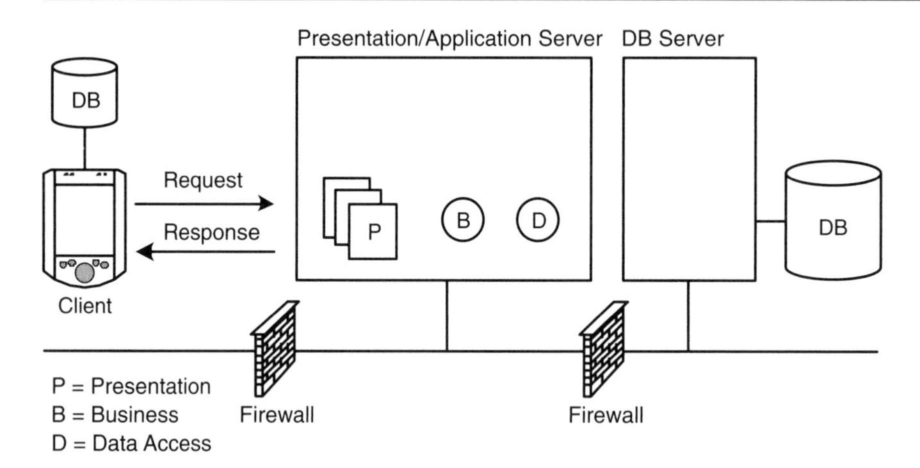

3.3.3 Three-Tier Architecture

A three-tier architecture can be developed so that the database, application, and presentation servers are split off from one another (see Figure 3–14). There are several pros and cons for doing so, as follows:

Pros

- Scalable
- Secured behind firewalls and zones
- Allows database server specialization

Cons

- Overkill
- More difficult to develop
- More difficult to manage
- More expensive

Splitting off the database allows the database server to become a more specialized server. Splitting off the presentation and application servers also allows for specialization of those servers. This allows you enormous scalability potential. It is also possible to secure the application and servers behind firewalls and zones since the presentation servers can be placed in the DMZ and application and database servers can be placed in the Intranet Zone.

Figure 3–14 Three-tier architecture

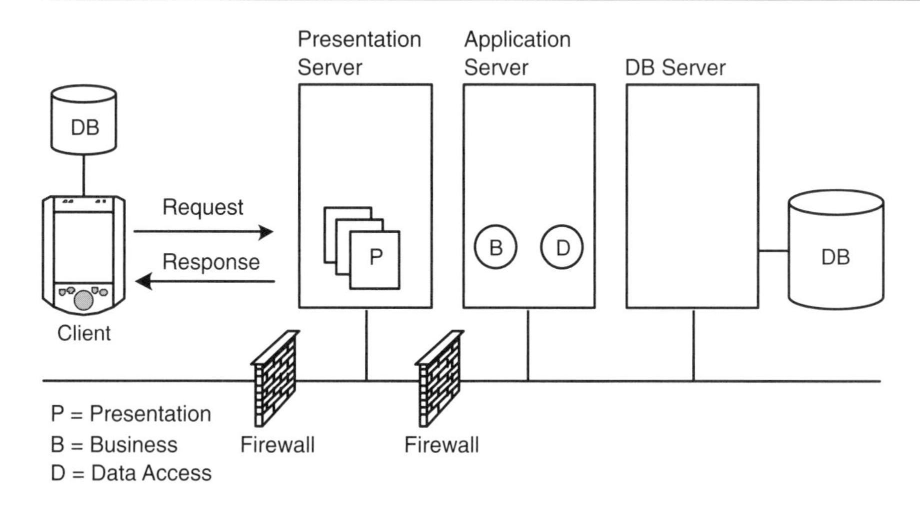

3.4 CONNECTION TYPES

As discussed in Chapter 1, mobile devices typically operate in one of three modes: always connected, partially connected, and never connected (see Figure 3–15). These modes are described in more detail in the sections that follow.

3.4.1 Always Connected

A mobile device, such as a cellular telephone or RIM device, normally operates in an always connected mode. In fact, RIM coined the phrase "always on, always connected."

An enterprise might have a wireless network and set of applications and servers that allow employees to connect and use their mobile devices while on company premises. Mobile devices, such as PDAs, Tablet PCs, and Laptop PCs, then essentially become extensions of the existing applications and infrastructure, permitting users the ability to always be connected to the applications while freely moving about the office.

3.4.2 Partially Connected

While the vision has been that mobile devices should always be connected, there are many scenarios where the mobile device is actually out of contact for extended periods of time.

Ironically, to a mobile application developer, this is where things are most interesting. For example, a mobile office worker might periodically connect to a server at the office to obtain email, contact information, or tasks to be done. The worker then disconnects the mobile device and carries out his/her normal tasks away from the office, during which time he/she might refer to the downloaded information. The user might also update the information locally on his/her mobile device before reconnecting at a later time to resynchronize the mobile device with the server.

3.4.3 Never Connected

There are also several mobile devices that never connect to back-end systems, such as certain gaming devices. While we have included this section for completeness, we will not be discussing these devices in this book.

Figure 3–15 Connection types

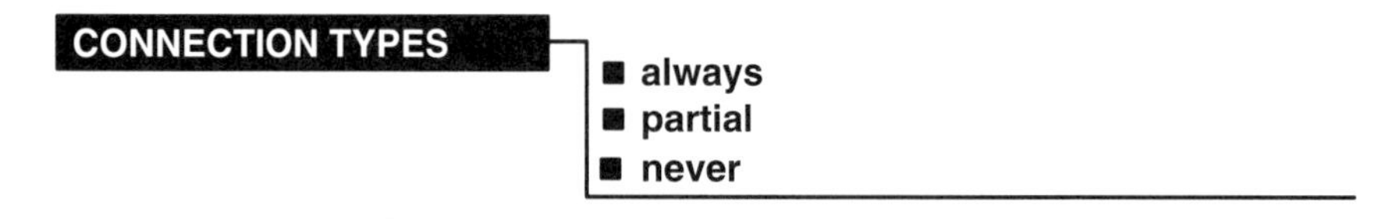

3.5 SYNCHRONIZATION

The connection type affects the way in which you can synchronize data between the mobile device and back-end systems. Synchronization is possible in two ways: continuously or through a store-and-forward method (see Figure 3–16). In the following sections, we will discuss these methods in more detail.

3.5.1 Continuous Communication

When the connectivity between the client and server is continuous, the synchronization of data between client and server is continuous and can be achieved through synchronous or asynchronous means (see Figure 3–17).

Figure 3–16 Synchronization methods

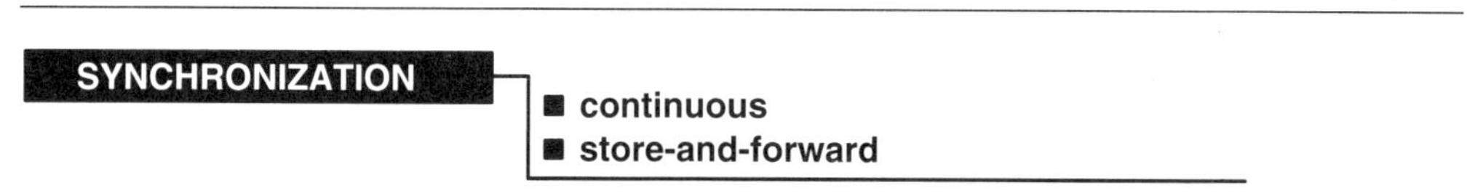

Figure 3–17 Synchronous and asynchronous communication

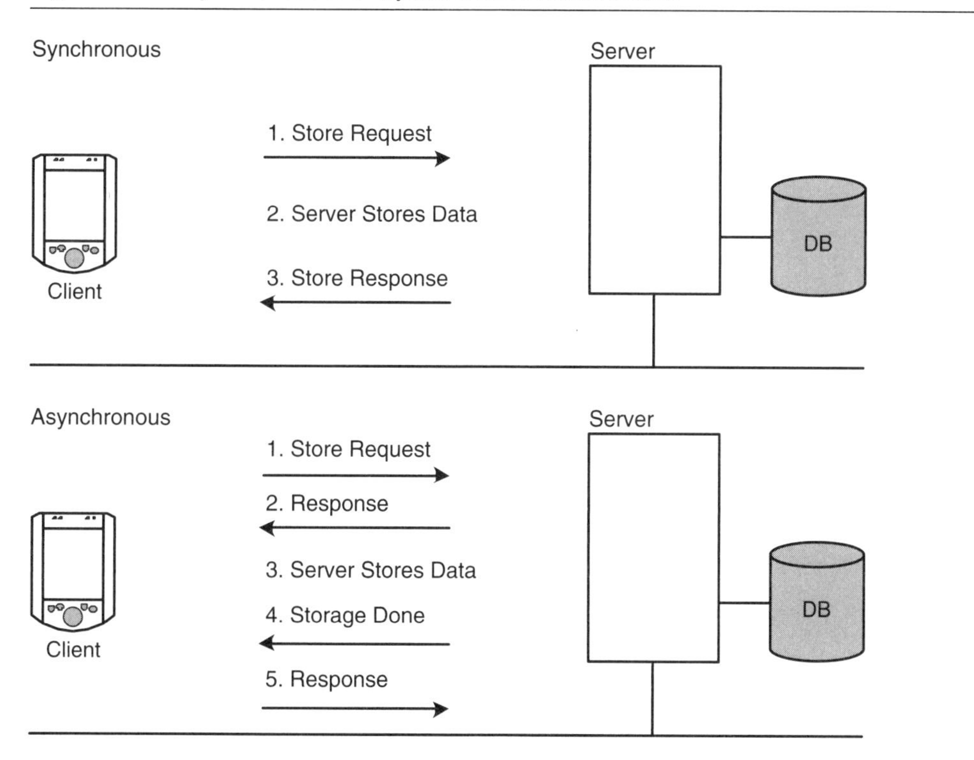

Synchronous communication occurs when a request to store data is sent to the server followed by the data to be stored. The data is then placed in a storage area, such as a database, on the server. In synchronous communication, all data is completely stored before the server acknowledges receipt of the data and frees up the client user interface.

Asynchronous communication occurs when a request to store data is sent to the server followed by the data to be stored. The data is then placed in a storage area, such as a database, on the server. In asynchronous communication, however, the data does not have to be completely stored before the server acknowledges the client. Indeed, the server typically acknowledges the request immediately and only subsequently carries out the store request. Subsequently, when the store request is actually complete, the server will initiate a conversation to tell the client it is done.

3.5.2 Store-and-Forward Synchronization

When connectivity between a client and server cannot be guaranteed, it is still possible to store and transmit information safely using a method called "store-and-forward."

Suppose, for example, that a mobile user wishes to enter data while his/her mobile device is not connected to a server. A mobile client application can initially store the data in a local data-

Figure 3–18 Store and forward

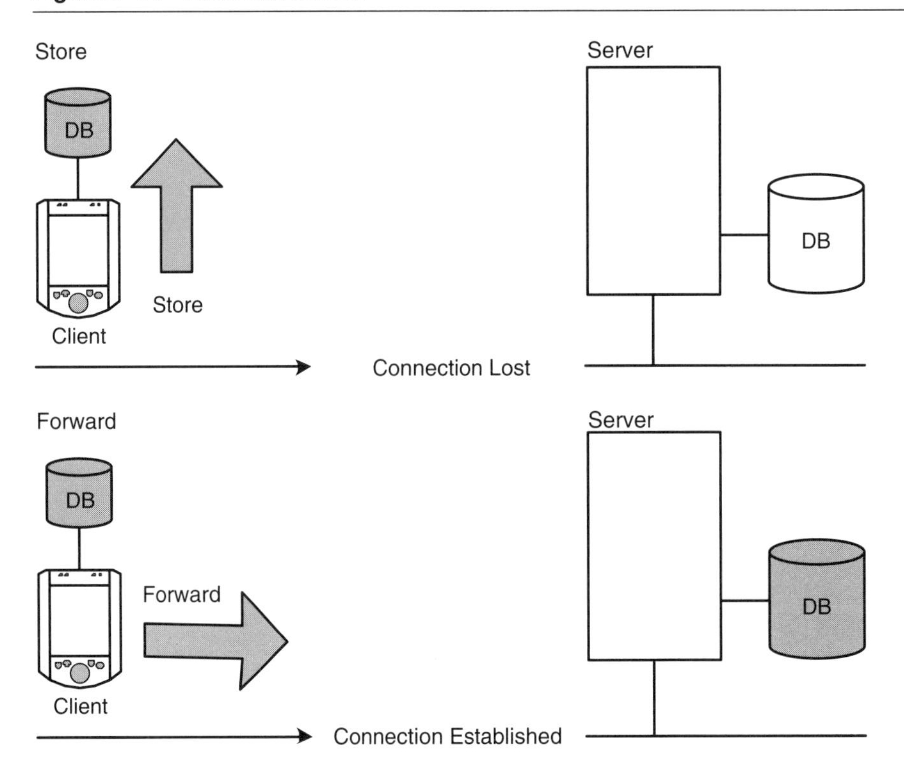

base. Later, when a connection has been established, the mobile application will forward the data from the local database to the database on the server (see Figure 3–18).

Store-and-forward is a powerful method that allows mobile users the ability to work even when they are not connected to a server. It is important to note, however, that if you permit mobile users to store data in a local database in this manner, you must also ensure data integrity when the data is synchronized with the server database, since other users may be adding or modifying possibly conflicting data on their mobile devices.

3.6 INTERESTING ARCHITECTURAL PATTERNS

In the following sections, we describe some of the patterns of small-, medium-, and large-scale mobile application architectures.

3.6.1 Pattern Matrix

If we assume there are four possible client layers, three possible server tiers, and three connectivity types, there are 36 possible combinations in total. This can be represented in tabular form (see Table 3–1).

However, not all of these combinations are particularly useful or viable. For example, if a mobile client device is never connected to a back-end server, this is not a useful architecture in our case. Thus, 12 of the 36 combinations are not viable, leaving just 24 patterns.

Currently, the "partially connected" patterns are probably the most prevalent since connectivity cannot always be guaranteed. In the future, the "always connected" patterns can be expected to be much more prevalent.

3.6.2 Zero-Layer, Three-Tier, Always Connected Architecture

In Figure 3–19, we present a simple mobile architecture. The mobile client has zero application code layers on it, which means it is a thin client. The server holds all the application code and it is organized in a three-tier architecture.

The Presentation Tier has application code that is able to render pages to a mobile device such as a Pocket PC. The pages are ordinary web pages (e.g., ASP.NET, JSP, HTML) and are viewable through the use of a web browser such as Microsoft Pocket Internet Explorer.

The Presentation Tier also communicates with business and data access objects on the Application and Database Tiers. Typically, data may be read from the database and written back to it during an update.

This architecture is very simple because the mobile client is assumed to always be connected to the server. Thus, there is no provision for storing application data on the mobile device. If the mobile device becomes disconnected, it will not be able to obtain up-to-date information until the connection is re-established.

Table 3–1 Pattern Matrix

Client	Server	Connectivity	Comments
0	1	Always	
0	1	Partial	
0	1	Never	Not Viable
0	2	Always	
0	2	Partial	
0	2	Never	Not Viable
0	3	Always	
0	3	Partial	
0	3	Never	Not Viable
1	1	Always	
1	1	Partial	
1	1	Never	Not Viable
1	2	Always	
1	2	Partial	
1	2	Never	Not Viable
1	3	Always	
1	3	Partial	
1	3	Never	Not Viable
2	1	Always	
2	1	Partial	
2	1	Never	Not Viable
2	2	Always	
2	2	Partial	
2	2	Never	Not Viable
2	3	Always	
2	3	Partial	
2	3	Never	Not Viable
3	1	Always	
3	1	Partial	
3	1	Never	Not Viable
3	2	Always	
3	2	Partial	
3	2	Never	Not Viable
3	3	Always	
3	3	Partial	
3	3	Never	Not Viable

Figure 3–19 Zero-layer, three-tier, always connected architecture

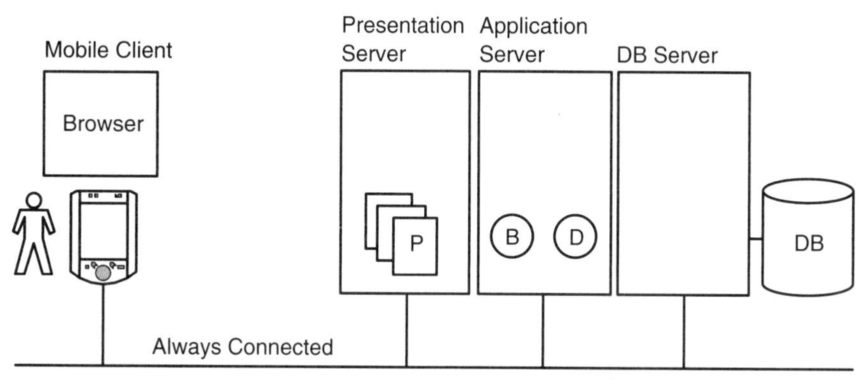

3.6.3 Three-Layer, Three-Tier, Partially Connected Architecture

In Figure 3–20, we present a more complex mobile architecture. The mobile client has three application code layers, which means it is a fat client. The server also holds application code and is organized in a three-tier architecture.

The mobile client has a complete standalone application that is able to read and write user-entered data to a local database during periods when it is not connected to the server. When connectivity has been re-established, the data can be retrieved from the local database and uploaded to the server using a store-and-forward mechanism.

Figure 3–20 Three layer, three-tier, partially connected architecture

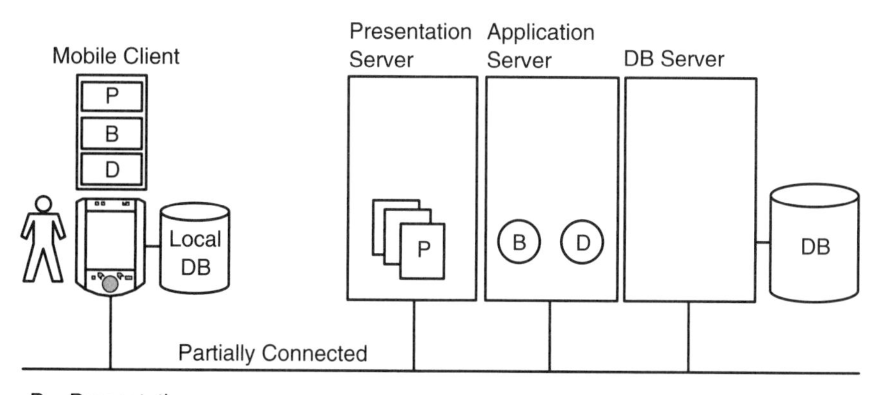

3.7 GOOD ARCHITECTURAL DESIGN TENETS

In the following sections, we describe some of the tenets of good architectural design. In practice, it may not be possible to achieve all of the tenets. Nonetheless, you will find that many of the best mobile application architectures meet many of these tenets.

3.7.1 Requirements

The architectural design must address the business, functional, and user requirements. Without conforming to the requirements, you will not be able to satisfy financial, functional, or usability success criteria.

3.7.2 Technology Independence

In an ideal world, you should develop mobile applications that are as device and platform-independent as possible. This is not always easy or possible, but good applications tend to be written so that they can run on many devices and platforms.

In practice, however, most applications will probably fall short of these paradigms. In all likelihood, you will need to select a preferred device and platform and write your application accordingly. Thus, you will almost certainly have to choose a mobile device such as the HP iPAQ, which uses Microsoft Windows Mobile 2003, or an HP Tablet PC running Microsoft Windows XP Tablet Edition. Alternatively, you may choose a Palm Pilot running Palm OS. Each device and platform has different characteristics that your application must take into account.

3.7.3 High Performance and Availability

The architecture must typically have excellent performance during normal and spike periods in resource demand. For example, for an e-business brokerage trading site, this might be on a heavy stock trading day. If people can be expected to use the site at any given time, the architecture must also be highly available.

3.7.4 Scalability

The architecture must be scalable to accommodate possibly large increases in the number of users, applications, and functionality. The architecture must typically be designed to easily allow for both horizontal (adding more servers) and vertical (adding faster servers) scaling without adversely affecting any existing applications.

3.7.5 User System Requirements

The architecture should typically handle the widest range and number of users possible. For example, a web application with large graphics to be displayed on a Pocket PC may look beautiful, but if users only possess low-speed modem lines, the performance may not be satisfactory. Thus, the full range of users must be kept in mind, including those with high and low performance systems.

3.8 SUMMARY

Mobile application architectures can be modeled conveniently in terms of client-server architectures. Clients can be always connected, partially connected, or never connected to a server. The code on a client or server can be layered. A client can have zero to three layers while a server has one to three layers.

Mobile client devices typically can contain thin clients or fat clients, or they can be developed so that they can host web pages. The server architecture can have one to three tiers. There are pros and cons associated with developing architectures with different tiers.

If the client is always connected to the server, it is possible to build a thin client and server architecture with no code on the client. If the client is partially connected, a fat client may be needed. Alternatively, a hosted set of web pages may be employed. In any case, good architectural design tenets, such as availability and scalability, should be followed.

CHAPTER 4

Mobile Infrastructure

For I dipped into the future, far as human eye could see,
saw the vision of the world, and all the wonder that would be...

—Alfred, Lord Tennyson

In this chapter we consider the mobile infrastructure from a mobile application developer's perspective. We start by describing the characteristics of several mobile device types, including pagers, cellular telephones, PDAs, Tablet PCs, and Laptop PCs. We then describe a mobile device's basic components and capabilities from a hardware perspective. Finally, we discuss several connection methods that exist to establish a connection between a mobile device and a back-end network, server, and application.

Understanding mobile infrastructure makes you a more well-rounded application developer and allows you to make more informed decisions about the infrastructure upon which your application must run.

4.1 MOBILE DEVICE TYPES

A wide range of mobile device types exist on the market today, aimed at both enterprise users and general consumers. The intrinsic capabilities, functions, portability, and cost of each of these mobile devices vary greatly. As a result, there are many ways to classify the device types. One that is helpful to consider is shown in Figure 4–1.

Figure 4–1 Mobile device types

MOBILE DEVICE TYPES

- **pagers/RIM devices**
- **cellular telephones**
- **pda devices**
- **tablet pc**
- **laptop pc**

In the following sections, we will discuss each mobile device type in more detail. We will start with a general description of each mobile device type followed by a description of their features. We will then discuss some of the considerations you need to make when purchasing a particular mobile device.

4.1.1 Pagers/RIM Devices

Pagers are used mainly by people who have to be reachable for their expertise or ability to do something. Common examples are medical service personnel (e.g., doctors and nurses who are on call), computer operations personnel, and traders.

To page someone, a person telephones a paging service and inputs the number of the person to be paged, followed by a telephone number or a short message indicating where the person telephoning can be reached. This telephone number or message is sent to the paged person who responds by calling back. More recent systems implement two-way messaging, where the paged person can send back a short response. In addition, many systems now support the delivery of wireless email.

There are many service providers in this arena, including:

- Cingular
- Nextel
- Research In Motion (RIM)
- Skytel

RIM, for example, makes two basic product device lines with different capabilities, namely:

- BlackBerry devices
- RIM devices

RIM devices now offer multiple services in addition to basic paging services. These include:

- Wireless email
- Simple Messaging Service (SMS) (e.g., text messages)
- Information (e.g., quotes, weather)
- Custom applications (e.g., Java)

RIM devices have a proprietary operating system and use data-only networks such as Mobitex and DataTAC wireless data protocols, while RIM BlackBerry devices use Global System for Mobile Communications (GSM)/General Packet Radio Service (GPRS) networks. These networks have extensive coverage and it is thus possible to send and receive pages and email reliably.

Indeed, one of the main reasons pagers are so successful is their reliability. Even today, if you want to guarantee that a message is delivered, a pager is your safest choice. RIM, for example, is the company that coined the slogan "always on, always connected" in reference to their devices' normal operation mode.

Traditionally, pagers have not had many applications running on them. Recently, however, the capabilities of these mobile devices have increased to such an extent that they are fast becoming PDAs in their own right. For example, the latest RIM BlackBerry now allows you to view Microsoft Word, Microsoft Excel, Microsoft PowerPoint, Adobe PDF, and Corel WordPerfect document attachments in email.

Therefore, while we still consider RIM devices to be two-way pagers, a case can be made that top-of-the-line RIM devices are becoming as sophisticated as PDAs.

4.1.2 Cellular Telephones

We initially hesitated about devoting a complete section to cellular telephones. After all, we reasoned, who doesn't know what a cellular telephone is and what it can do?

However, the more we looked into it, the more we realized that even we didn't realize just how far the cellular telephone has come in the last few years in terms of functionality. As a result, we felt it necessary to talk at least a little bit about this mobile device.

A very wide range of cellular telephones exist on the market today and are aimed at all types of mobile users from enterprise users to young and adult consumers. There are a large number of cellular telephone vendors, including:

- AudioVox
- LG
- Motorola
- Nokia
- Orange
- Samsung
- Sony Ericsson
- T-Mobile

Their products can be used with a variety of service providers, including:

- AT&T
- Cingular
- Motorola
- Nextel
- NTT DoCoMo
- Orange
- Sprint
- T-Mobile
- Verizon

Cellular telephones now offer multiple services and functions in addition to basic phone services, including:

- Wireless email
- Multimedia Messaging Service (MMS) (e.g., text messages with pictures and sounds)
- SMS (e.g., text messages)
- Enhanced Messaging Service (EMS) (e.g., voice messages)
- FM radio
- Camera
- Flashlight
- Internet access (using WAP over GPRS)
- Games (e.g., BREW)
- Information (e.g., quotes, weather, location services)
- Custom applications (e.g., Java)

Some cellular telephones still have monochromatic screens that can display only five to six lines of text and do not have a complete keyboard. However, many advanced cellular telephones now have small color 128×128 pixel screens and complete keyboards for input. In spite of these and other innovative developments, the cellular telephone still has limitations compared to other mobile devices, most notably in its user interface.

4.1.3 PDAs

The term "Personal Digital Assistant" (PDA) was first coined by Apple Computers, Inc. in reference to their Newton MessagePad device, which allowed users to write data to the device. Since then, several other vendors have released increasingly powerful lines of PDAs. Today, there are many PDA vendors, including:

- HP (Microsoft Windows OS)
- PalmOne (Palm OS)
- Sony (Palm OS)
- Toshiba (Microsoft Windows OS)

Operating systems from PalmSource and Microsoft currently dominate the PDA market. PDA devices that run the Microsoft operating system are often called Pocket PCs, ostensibly because they fit into a shirt pocket.

Originally, the PDA was intended to be an electronic version of a "personal organizer." Thus, it had a clock, a calendar for business appointments, a task list, and a telephone directory for contact information as part of its core functionality.

However, with the introduction of more powerful CPUs, operating systems, and memory, today's PDA has many more functions, including:

- Email
- Internet access
- Games
- Information (e.g., quotes, weather, location services)
- Custom applications (e.g., .NET, Java)

A PDA also typically contains a local copy of the information stored on a user's desktop PC. This allows a user to work un-tethered. When an update is needed, the PDA is cradled so that it can receive updates from the desktop PC and also send any changes a user has made back to the desktop PC. For example, a user might write an email on a PDA and store it locally. Upon cradling, synchronization software detects the new email and uploads it to the user's desktop for delivery.

Data entry through the use of a stylus and touch-screen along with handwriting recognition has been a feature of PDAs since their inception. In addition, the ability to power up the device very rapidly (in one second or so) has also been a very useful feature.

While these features have paved the way for some highly innovative uses, users have found writing on a small PDA difficult and the device's handwriting recognition mechanism weak. In addition, users have also found tapping away on a small-screen image of a virtual keyboard for prolonged periods of time to be very cumbersome. The lack of permanent storage and the relative ease with which data can be lost have also been concerns, although both can be alleviated somewhat through the use of an external storage mechanism, such as a Compact Flash (CF) or Secure Digital (SD) memory card.

4.1.4 Tablet PCs

A Tablet PC is a general purpose mobile computer with a large, integrated, interactive screen. Perhaps its most compelling feature is its ability to allow users to write comfortably and directly onto the screen using a pen.

A range of Tablet PCs exist on the market today. Some of the major Tablet PC vendors include:

- Acer
- Fujitsu
- HP
- NEC
- Toshiba

As a general purpose mobile computer, a Tablet PC has nearly all the functions that a desktop computer has, including:

- Email
- Internet access
- Games
- Information (e.g., quotes, weather, location services)
- Office applications (e.g., word processing, spreadsheet processing, presentation generation)
- Multimedia (e.g., digital video, pictures, audio)
- Custom applications (e.g., .NET, Java)

Tablet PCs have actually been around since 1989, the year GRiD Computing released the GRidPad, a portable computer with a tablet design. Since then, several companies have periodically produced increasingly sophisticated and powerful tablet devices, including the GO PenPoint in 1991, the Apple Newton MessagePad in 1993, the Palm Pilot in 1996, the Vadem Clio in 1997, and the HP iPAQ in 1999. At the end of 2002, HP released the TC1000 Tablet PC, which runs Microsoft XP Tablet PC Edition.

Notice that a PDA device can be considered a type of Tablet PC and, indeed, from a writing perspective, it is. However, unlike a PDA, a Tablet PC runs a more powerful and fully featured operating system. It also has a hard disk drive. Unfortunately, these features also mean that you cannot boot up a Tablet PC nearly as rapidly as a PDA (it takes a few minutes versus a second or so).

Nonetheless, the Tablet PC's large touch-screen, pen, keyboard, and sophisticated handwriting recognition and drawing capability have raised hopes that it might become the mobile device of choice for enterprise workers who not only have to be mobile but also have to read, write, and type a great deal. For example, doctors or nurses on rounds who have to enter patient information might use a Tablet PC instead of paper and a clipboard.

4.1.5 Laptop PCs

A Laptop PC is a general purpose mobile computer with a large, integrated screen that uses a mouse and a keyboard as typical input devices.

The Laptop PC is intended to be a full-featured mobile version of a desktop computer. For example, it is perfectly possible to run a full operating system and application development software such as Microsoft Windows 2003 Server and Microsoft Windows Visual Studio.NET on a Laptop PC. Unlike a PDA or a Tablet PC, however, a Laptop PC does not have a touch-screen or the capability for stylus/pen input.

A very wide range of Laptop PCs currently exist. Some of the major vendors of Laptop PCs are:

- Apple
- Dell

- Fujitsu
- HP
- IBM
- Sony

As a general purpose mobile computer, a Laptop PC has all the functions that a desktop computer has, including:

- Email
- Internet access
- Games
- Information (e.g., quotes, weather, location services)
- Office applications (e.g., word processing, spreadsheet processing, presentation generation)
- Multimedia (e.g., digital video, pictures, audio)
- Custom applications (e.g., .NET, Java)

A Laptop PC mainly differs from a desktop computer in terms of its hardware capacity and portability. A Laptop PC's CPU speed, memory space, disk space, and screen size are typically less than that of a desktop computer. However, it is also probably the largest mobile device that can generally be considered portable today.

4.1.6 Hybrids

The primary functions of certain mobile devices have been combined into new *hybrid* mobile devices that have overlapping functionality.

Currently, high-end RIM devices and cellular telephones almost rival PDAs by providing many PDA services (including email and access to the Internet) on top of paging and cellular telephone services.

In addition, "Smartphones" that integrate PDA functionality with cellular telephone functionality have also recently become available. These devices give users the ability to check email, listen to music, edit files, and browse the Internet, as well as make and receive phone calls. Samsung and Motorola have released Smartphones that run Microsoft Windows Mobile 2003 Phone Edition.

While many mobile device functions can be combined, there are also tradeoffs in doing so. For example, a Smartphone that is the size of a traditional PDA may be too wide to hold comfortably to your ear as a telephone for long periods of time. Conversely, a Smartphone that is the size of a traditional cellular telephone may have a screen that is too small to easily perform tasks like checking email. It is also difficult to read or write to a PDA while simultaneously listening on the telephone. (A wired retractable earpiece or wireless earpiece might be a handy little feature to alleviate this problem.)

There may be some convergence of technologies at the lower end of the spectrum, wherein RIM devices, cellular telephones, and PDAs converge into one device that is always on and connected while also having the powerful functionality offered by a PDA. At the higher end of the spectrum, we may also see Tablet PC-type devices with integrated telephony capabilities.

It will be interesting to see if an ideal is ever achieved. Our guess is that it is unlikely, simply because every user's needs are different. Nonetheless, many innovative and unusual combinations will undoubtedly be tried and some will become very popular if they suit a large number of people.

4.1.7 Capability and Cost Considerations

In the enterprise, your choice of mobile device type is also constrained by what your business wants to do and can afford to do. An important consideration you will have to take into account is what the primary function of the user's mobile device will be. As we've seen, not every mobile device does the same thing, nor is it necessarily capable of the same functionality.

For example, if the user needs to type a great deal, a cellular telephone or a PDA may not be as suitable as a Tablet PC or a Laptop PC. However, if the user only uses a mobile device for simple calendaring functions or email, a RIM device may be a sufficient, cost-effective alternative to a high-end PDA.

The capabilities of a mobile device are constrained both by its size and cost. Generally, therefore, the greater the capability and functionality of a mobile device, the larger and more costly the mobile device will be (see Figure 4–2). Note that this is not a hard and fast rule, and it is not difficult to find exceptions to it. For example, fully featured PDAs and Tablet PCs can be more expensive than Laptop PCs.

Figure 4–2 Mobile device Capability/Functionality vs. Size/Cost

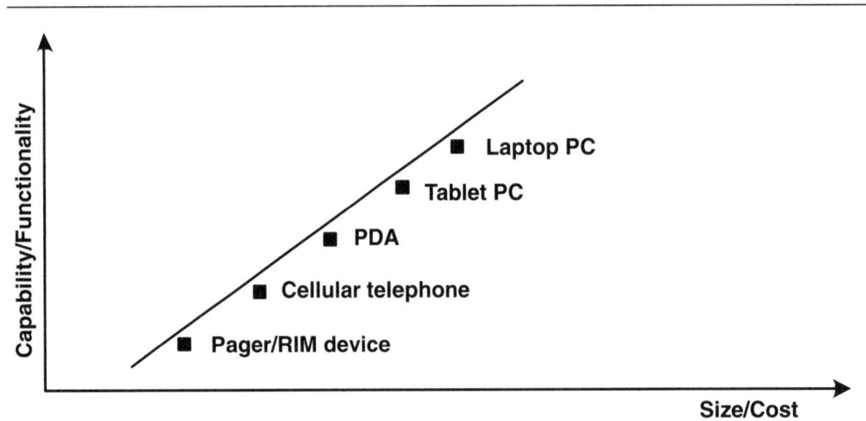

4.2 MOBILE DEVICE COMPONENTS

A mobile device contains many of a typical computer's components, including a CPU, operating system, memory, disk, batteries and power, connection ports, screen, keyboard, mouse, stylus and pen, and a set of peripherals (see Figure 4–3). While the operating system is not hardware, it is nonetheless useful to consider it here because the operating system has a powerful influence on the mobile device's features and capabilities.

As a rule of thumb, we can generally say that when you are attempting to judge the characteristics of mobile devices from a hardware perspective, you want more of everything you can afford. Simply put, you are looking for "bang for your buck."

Ideally, therefore, you want a faster CPU, more memory, more disk space, longer lasting batteries or more efficient power usage, more connection options, more capability and functionality, and greater convenience in terms of portability and usability.

Naturally, each of these items comes at a price. Very fast CPUs are expensive. They can run at high temperatures and require cooling mechanisms that take up valuable space. Disk drives and batteries are heavy and reduce your portability. Thus, it is unlikely that you will be able to have every feature even with the enormous range of prices and selections that are available.

As a result, you will probably need to make tradeoffs depending on your specific circumstances and what features are most important.

In the following sections, we will describe the components of a mobile device in more detail. We will also discuss the tradeoffs involved in finding an ideal device for users in terms of capability, suitability, portability, and price.

4.2.1 CPU

A CPU is of critical importance to a mobile device's operation. A CPU has a maximum clock speed that determines the frequency at which it can fetch and execute instructions. This is one of several key indicators of the mobile device's overall performance.

Figure 4–3 Mobile device components

DEVICE COMPONENTS

- **CPU**
- **operating system**
- **memory**
- **disk**
- **batteries and power**
- **connection ports**
- **screen**
- **keyboard**
- **mouse, stylus, pen, and voice**
- **peripherals**

In general, mobile devices with the fastest CPUs are more expensive. In part, this is because it is expensive to develop and manufacture advanced CPUs. However, you also pay more for the manufacturer's brand name. Lesser known manufacturers may be able to provide CPUs with comparable speeds that are less expensive. This is, in turn, counterbalanced by the risks associated with purchasing a CPU from a company that you may not know.

In Table 4–1, we show typical CPU speeds for several mobile device types. The data should not be construed as highly accurate since they are only intended to illustrate the differences in device capabilities.

Most CPU microprocessors on the market today are made by several major manufacturers, including:

- AMD
- Intel
- Motorola
- Texas Instruments
- Transmeta

Intel has a very broad range of microprocessor families that are used for many purposes. For example, Intel's 386 microprocessors are still used in RIM BlackBerry devices, while Intel's XScale microprocessors are used in PDAs. In addition, the Intel Pentium III, IIIm, 4, 4m, and M microprocessor lines are widely used in Tablet PCs and Laptop PCs.

AMD also makes a broad range of microprocessors, including the AMD Athlon XP-M, Mobile AMD Athlon 4, and Mobile AMD Duron, which are widely used in Laptop PCs.

Another manufacturer of interest is Transmeta, whose Crusoe line of microprocessors is widely used in HP Tablet PCs.

Texas Instruments is another large manufacturer with a wide portfolio of microprocessors. One microprocessor family of particular interest is the OMAP family, which is widely used in devices ranging from Nokia SmartPhones running Symbian OS to PalmOne PDAs running Palm OS.

ARM develops and provides 16-bit and 32-bit embedded Reduced Instruction Set Computer (RISC) microprocessors for mobile devices, particularly cellular telephones and PDAs. ARM also licenses its RISC processors, peripherals, and system chip designs to leading interna-

Table 4–1 Typical CPU Speeds

Mobile Device Type	Typical CPU Speeds
Pager/RIM device	20–40 MHz based on Intel 386
Cellular telephone	-
PDA	200–400 MHz based on Intel XScale
Tablet PC	1 GHz
Laptop PC	1–2 GHz

tional electronics companies such as Intel, Nokia, and Motorola. For example, the Intel Strong-ARM chip is used in PDAs made by HP and other Pocket PC manufacturers. Nokia also uses ARM technology in some of its cellular telephones, while some of Texas Instruments' OMAP processors contain an ARM microprocessor core used for PDAs.

4.2.2 Operating System

When you implement a mobile application, it is important to consider the operating system of the mobile device. The operating system affects the language, tools, and technologies that you use to develop your mobile application, as well as your ability to support and maintain the application.

In Table 4–2, we show some of the operating systems for several mobile device types. The data should not be construed as highly accurate since they are only intended to illustrate the differences in device capabilities.

Several vendors make operating systems for mobile devices, including:

- Microsoft
- PalmSource
- Psion
- RIM
- Symbian

Currently, at the lower end of the mobile device type spectrum, RIM's proprietary operating system is used on RIM devices, while cellular telephones use Psion EPOC or Symbian OS. Palm OS and Microsoft Windows operating systems dominate the PDA market, while at the higher end of the mobile device spectrum, Microsoft Windows operating systems dominate the Tablet PCs and Laptop PC market.

4.2.3 Memory

A mobile device's CPU fetches and executes instructions and stores temporary information in memory. Ideally, you want to have as much memory as possible for performance purposes. However, the need for more memory can be constrained by cost.

Table 4–2 Typical Operating Systems

Mobile Device Type	Typical Operating Systems
Pager/RIM device	RIM OS
Cellular telephone	Windows Mobile 2003 Phone Edition, Psion EPOC, Symbian OS
PDA	Windows Mobile 2003, Palm OS
Tablet PC	Windows XP Tablet Edition
Laptop PC	Windows XP, Linux, Mac OS

Table 4–3 Typical Memory Size

Mobile Device Type	Typical Memory Size
Pager/RIM device	4 MB–16 MB Flash ROM
Cellular telephone	56 MB
PDA	64 MB SDRAM; 48 MB Flash ROM
Tablet PC	256 MB 1 DDR SDRAM
Laptop PC	1 GB

In Table 4–3, we show some typical memory sizes for several mobile device types. The data should not be construed as highly accurate since they are only intended to illustrate the differences in device capabilities.

Memory in mobile devices such as PDAs is particularly important because these devices do not have a hard disk drive and rely upon memory for their data storage. This has both positive and negative ramifications. One positive ramification is that starting up a mobile device is much quicker than an ordinary PC (e.g., a few seconds versus a few minutes). A negative ramification is that there is no permanent data storage for some mobile devices.

4.2.4 Disk

A mobile device may or may not have a disk drive for permanent data storage. In Table 4–4, we show some typical disk sizes for several mobile device types. The data should not be construed as highly accurate since they are only intended to illustrate the differences in device capabilities.

It is important to realize that not all mobile devices have hard disk drives on which to store data. For example, PDAs such as Pocket PCs typically do not have a hard disk drive. The lack of a hard disk drive on PDAs means you cannot rely on the storage capability of those mobile devices to the same extent that you might for a desktop or Laptop PC.

Thus, data storage on mobile devices such as PDAs can be somewhat transient in nature. Generally, data should be uploaded to a reliable system as soon as it is possible and convenient.

Table 4–4 Typical Disk Size

Mobile Device Type	Typical Disk Size
Pager/RIM device	-
Cellular telephone	-
PDA	-
Tablet PC	30 GB
Laptop PC	30–60 GB

4.2.5 Batteries and Power

A mobile device is highly dependent upon its batteries for power simply because the device wouldn't be considered highly mobile if it constantly had to be wired to the main power source.

Battery life varies widely, depending on the mobile device type, the mobile device's usage, and the actual battery technology itself. In Table 4–5, we show battery duration times for several mobile device types. The data should not be construed as highly accurate since they are only intended to illustrate the differences in device capabilities.

Pagers and RIM devices use batteries very efficiently and typically do not drain power rapidly. As a result, even though RIM devices are always powered on, simple AA batteries can be used for weeks before they need to be replaced.

Cellular telephone batteries currently can last for several days with non-continuous use before needing to be recharged.

PDAs, Tablet PCs, and Laptop PCs also use rechargeable batteries. However, because these devices all demand a great deal of power with continuous use, they typically need to be recharged every few hours.

Up-to-date mobile devices use efficient lithium ion batteries. In the near future, lithium polymer batteries may be used. Lithium polymer batteries are handy because they are pliable and can be twisted to be packed into odd spaces within a mobile device.

In the future, fuel cells may also come into use. These are highly efficient energy sources and are more environmentally friendly than today's batteries. However, fuel cells probably will not see widespread availability and use for some years.

4.2.6 Connection Ports

A mobile device typically has a number of connection ports cunningly hidden in all manner of places on the device or through an attached sleeve.

In Table 4–6, we show the speeds of the various connection ports. The data should not be construed as highly accurate since they are only intended to illustrate the differences in port capabilities.

Larger mobile devices, such as Tablet PCs and Laptop PCs, will have nearly all the aforementioned ports. However, smaller mobile devices, such as RIM devices, cellular telephones, and

Table 4–5 Typical Battery Duration (Under Average Use)

Mobile Device Type	Typical Battery Duration
Pager/RIM device	Weeks
Cellular telephone	Days
PDA	Hours/Days
Tablet PC	Hours
Laptop PC	Hours

Table 4–6 Connection Port Speeds

Port	Typical Speed
Bluetooth	56–721 Kbps
Firewire	400 Mbps
Infrared	4 Mbps
Parallel	50–100 Kbps
Serial COM	115–460 Kbps
USB 1.1	1.5–12 Mbps
USB 2.0	1.5–480 Mbps
RJ-11	9.6–56.6 Kbps
RJ-45	10–100 Mbps
PC Card Slots	10–100 Mbps

PDAs may or may not have these ports. When smaller mobile devices do not have these ports, a sleeve that is externally attached to the device often adds this capability.

Each port has a slightly different purpose because of its fundamental characteristics. For example, the wireless infrared port is often used to synchronize devices such as a Pocket PC, a Laptop PC, and a desktop computer. It has a very short range and requires devices to have line of sight between one another.

The Bluetooth port is a wireless port that is often used to synchronize a mobile device such as a Pocket PC to a host computer. It is a small radio module built into a mobile device that communicates with another device by radio waves. It also has a relatively short range, but does not require that devices have line-of-sight between one another.

The wired USB port can be used as a mouse port and also for attaching to printers and external drives. It is also used as a synchronization port between devices such as a Pocket PC and a host computer.

The Serial COM port is a wired port traditionally used as the mouse port. Today, it is also used to attach to printers and external drives.

The Parallel port is a wired port used mainly for attaching to printers. It is also occasionally used as the port to attach to external mass storage devices (e.g., Iomega Zip disk drives).

The RJ-11 phone line port is for dialup. While it is slow compared to some of the high-speed network lines and requires that you be wired to a telephone line, it still has one big advantage: it works almost anywhere. In other words, wherever you have a telephone line, you can attach a mobile device to it.

The RJ-45 network line port is a high-speed port that allows wired connections to a network. Typically, this port is found in Tablet PCs and Laptop PCs but not in RIM devices, cellular telephones, or PDAs.

The PC Card slots are also high-speed ports that allow wired or wireless network cards to be used on the mobile device. This includes the Wireless LAN 802.11 card.

4.2.7 Screen

Most, if not all, mobile devices have flat screens. The screen and battery are the main contributors to a mobile device's overall weight. Thus, the larger the screen, the heavier—and therefore less mobile—the device.

In Table 4–7, we show screen sizes for several mobile device types. The data should not be construed as highly accurate since they are only intended to illustrate the differences in device capabilities.

Two types of screens are commonly available for mobile devices: flat screens and interactive touch-screens. PDAs and Tablet PCs generally have touch-screens, which allow you to use styluses and pens to make precise screen selections and write without having to move a mouse around.

Generally, if a user won't read or write extensively on a mobile device other than perhaps composing quick emails or short text notes, you may find a PDA that synchronizes with a larger computing device sufficient for the user's needs.

However, if a user has to carry out any significant amount of reading or typing, Tablet PCs and Laptop PCs are the only truly viable options at present. RIM devices, cellular telephones, and PDAs are just too small for prolonged use.

4.2.8 Keyboard

In spite of some changes to the QWERTY keyboard over the years, it has steadfastly remained a part of the office environment since its inception as a typewriter. A full standard size keyboard is 7" × 18", which is large enough for an adult to comfortably type on and small enough not to be too cumbersome.

In Table 4–8, we show keyboard sizes for several mobile device types. The data should not be construed as highly accurate since they are only intended to illustrate the differences in device capabilities.

There are keyboard options and accessories available—such as rolled-up or folding keyboards—which allow you the luxury of having a bigger keyboard for more extensive typing on a smaller mobile device. However, this means that you have an additional peripheral to carry around with you.

Table 4–7 Screen Size

Mobile Device Type	Typical Screen Size
Pager/RIM device	3"
Cellular telephone	1.5"
PDA	4" (Handheld PC up to 9.4")
Tablet PC	10.4"
Laptop PC	10.4"–17"

Table 4–8 Keyboard Size

Mobile Device Type	Typical Keyboard Size
Pager/RIM device	Miniature
Cellular telephone	Miniature
PDA	None to very small
Tablet PC	Medium
Laptop PC	Medium to full sized

If a user does an extensive amount of typing, Tablet PCs and Laptop PCs are the only truly viable options; RIM devices, cellular telephones, and PDAs are just too small.

4.2.9 Mouse, Stylus, Pen, and Voice

In addition to a keyboard, there are several other mechanisms for inputting information into a mobile device, namely:

- Mouse
- Stylus
- Pen
- Voice

RIM devices, cellular telephones, and PDAs typically do not have a mouse. Buttons or a stylus are used to click or select options instead. However, on Tablet PCs and Laptop PCs, a mouse is usually available either built into the device or available externally by plugging it into the serial mouse port.

The stylus and pen are essentially pointers for selecting items and for writing. They are widely used in touch-screen PDAs and Tablet PCs.

There are other input methods available, such as issuing instructions through voice commands but, presently, this is not a widely used mechanism.

4.2.10 Peripherals

There are numerous accessories and attachments for mobile devices that we have loosely grouped together as "peripherals." We use peripheral in the traditional sense of the word—as a separate piece of hardware—even though some peripherals are now an integral part of mobile devices. There are several types of peripherals, including:

- Printers
- Cameras
- Barcode scanners

- Biometric scanners
- Sleeves (and modem and/or network card insertions)
- Location scanners (e.g., GPS outdoors, Newbury Network's Wireless Detection System (WDS) indoors)

Most of these peripherals are hardware attachments to the mobile device, accompanied by an application or driver that is installed on the mobile device. Generally, while they extend the functionality of the mobile device, they can also make the mobile device less portable since they affect its size and weight.

When you are selecting a mobile device on behalf of a user, it is important to consider the wide range of peripherals available since they can significantly help the user do his/her job. For example, if you are writing a mobile application for factory users who are taking stock inventory, a laser or barcode scanner may be very helpful. Similarly, if you are writing a mobile application for a government department that issues citations, you may need to consider a light-weight mobile printer for printing out the ticket.

It is also important to consider cost. Some of these peripherals are expensive and can lead to doubling or tripling the price of your basic mobile device. For example, a Pocket PC equipped with a barcode scanner and printer, multiplied over many mobile users, can be very expensive.

4.3 CONNECTION METHODS

Mobile devices can use either a wired or wireless connection in order to connect to a network to send and receive information (see Figure 4–4). In the following sections, we will explore some of the wired and wireless mechanisms that can be used in more detail.

4.3.1 Wired

There are several wired mechanisms that a mobile device can use to connect to a network to send and receive information (see Figure 4–5). In the following sections, we will explore some of these mechanisms in more detail.

Direct network connection. Many office workers are accustomed to having direct network connections to a 10 or 100 Base-T Ethernet network. These networks typically provide a data rate of 10 Mbps or 100 Mbps.

Figure 4–4 Connection methods

Figure 4–5 Wired connection mechanisms

WIRED MECHANISMS

- **direct network connection**
- **cradling**
- **dialup**

Many workers now also have the ability to create a high-speed direct Internet network connection from their home through the use of DSL or cable modems. This allows for much faster connection speeds than using dialup (e.g., 10 Mbps rather than 56 Kbps). Once connected, home office workers or telecommuters can tunnel into their office network by creating a virtual private network (VPN) connection.

PDAs, Tablet PCs, and Laptops PCs can establish these connections. In order to do so, the mobile device must have a Network Interface Card (NIC). Most Tablet PCs and Laptop PCs today have these cards pre-installed. However, users with PDAs, such as a Pocket PC, will first need to get a sleeve for their device and then insert a network card into the sleeve.

Cradling. In order to transfer data between a PDA and a host computer, the user inserts the PDA into a cradle (which is connected by a synchronization cable to a host computer) and uses synchronization software to transfer data between the two devices. A cradle and synchronization cable are standard parts of a complete PDA package.

Standard synchronization software allows you to create a partnership between a PDA and a host computer, and specify which items you want to synchronize. For example, Microsoft ActiveSync is able to synchronize email, appointments, tasks, AvantGo news, and other items between the two devices. It will also upload changes made on the Pocket PC to the host computer and download changes made on the host computer to the Pocket PC. If there are any conflicts, it will prompt the user to resolve them.

Microsoft Windows Mobile 2003 also allows the Pocket PC to share the host computer Internet connection while it is cradled. This means that you can browse the web on your Pocket PC while it is cradled.

Dialup. Dialup is one of the oldest and best established methods for users to connect to a network. Dialup connections are available for use with many mobile devices, including PDAs, Tablet PCs, and Laptop PCs.

In order to dial up to a network, the mobile device must have a modem. This modem can be an internal device built into the computer or an external device on the user's desktop. Modems today are very small and have come a long way from the book-sized modems of yesteryear. Currently, the fastest dialup modems typically transfer data at 56 Kbps.

Many PDAs, such as a Pocket PC, can be inserted into a sleeve that allows the user to add peripherals. For example, a small card-shaped modem can be inserted into the sleeve, allowing a PDA to be plugged into a telephone line.

In addition to the necessary hardware, a dialup user must have an account either on a corporate network or with an Internet Service Provider (ISP) that allows dialup access. The user typically connects the mobile device to the telephone line using the modem, dials the telephone access number, and enters an ID and password to connect.

Today, however, many people have become accustomed to the high-speed Internet connections they experience at work and often desire these higher connection speeds from home. This can now be achieved through the use of DSL and cable modems.

However, DSL and cable modems are still somewhat expensive options. In addition, high-speed Internet access is not available in all parts of the United States, and so dialup is still the only choice for many workers and consumers who need Internet or office network access from home.

One major disadvantage to dialup is that it ties up a user's telephone line while he/she is connected to the network. As a result, many users (especially home office workers) lease a second telephone line so they can speak on the telephone using one telephone line while surfing the Internet on the other. For users who require this dual connectivity, the speed and convenience of DSL or a cable modem may outweigh the cost of maintaining two separate telephone lines.

4.3.2 Wireless

Certain mobile devices, such as cellular telephones, pagers, and RIM devices, communicate exclusively using wireless connections. Other mobile devices that may communicate wirelessly include PDAs, Tablet PCs, and Laptop PCs.

There are several wireless mechanisms that a mobile device can use to connect to a network to send and receive information (see Figure 4–6). In the following sections, we will explore these mechanisms in more detail.

Figure 4–6 Wireless connection mechanisms

WIRELESS MECHANISMS

- **cellular**
- **data networks**
- **bluetooth**
- **wireless LAN**
- **satellite networks**

Cellular. Cellular networks are made up of sets of contiguous radio coverage areas called *cells*. These cells can range in size from approximately 100 yards across to 20 miles across and are typically illustrated as hexagons (see Figure 4–7).

When mobile device users are within a cell, they are able to transmit and receive radio signals from one or a number of omni-directional antennas located at the center of each cell, and thereby utilize the mobile device's services.

When a user moves from one cell to another, automatic procedures are initiated by the network to hand off the user to a new cell. If the coverage is good, this operation is transparent to the user and the user can continue with his/her call or data transmission. If the coverage is less comprehensive, it is possible for a call to be dropped.

There are several network access methodologies that subscribe to the general cellular model:

1. Frequency Division Multiple Access (FDMA)—This system carries each call on a separate frequency. It is typically used in analog cellular networks such as the Advanced Mobile Phone Service (AMPS) network. The AMPS network runs in the 824 MHz to 894 MHz frequency range and each channel is 30 KHz wide.
2. Time Division Multiple Access (TDMA)—This system assigns each call a certain time slice on a given frequency. Each 30 KHz channel is divided into three time slots, so TDMA networks can handle three times more calls than FDMA. TDMA runs in the 800 MHz frequency band (IS-54) and the 1900 MHz frequency band (IS-136).
3. Global System for Mobile Communication (GSM)—This system uses TDMA as its access method but is incompatible with TDMA (IS-136). It is used for digital cellular and Personal Communications Service (PCS) systems. GSM is the international

Figure 4–7 Cells

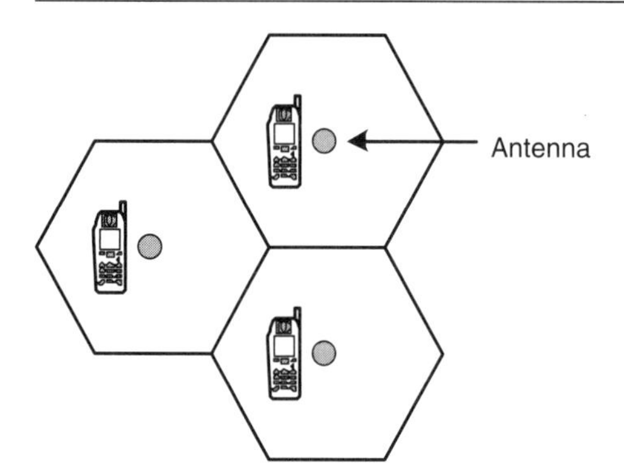

standard in Europe, Australia, and in much of Asia and Africa, where it operates in the 900 MHz and 1800 MHz bands. In the United States, however, GSM runs in the 850 MHz and 1900 MHz bands and is not compatible with the international standard.

4. Integrated Digital Enhanced Network (IDEN)—This system is run by Motorola and uses TDMA. It is used for digital cellular, text messaging, Internet access, and two-way digital radio.
5. Code Division Multiple Access (CDMA)—This system assigns each call a unique code. CDMA uses a spread spectrum to transmit each call over multiple frequencies in the band. CDMA operates in the 800 MHz and 1900 MHz bands. CDMA can handle eight to 10 times as many calls as the analog AMPS system.
6. Personal Communications Services (PCS)—This system is a wireless system similar to digital cellular, which bundles services like paging, caller ID, and email. PCS has smaller cells and operates in the 1850 MHz to 1990 MHz frequency band. PCS is based on TDMA but each channel is 200 KHz wide and divided into eight time slots.

Vendors that support AMPS, TDMA, and GSM include AT&T and Cingular, while CDMA is supported by Verizon and Sprint PCS.

There are several different standards for transmitting data over these cellular networks. The speed of data transmission varies based on the cellular access methodology and the specific transmission standard. For example, some typical data transmission speeds for GSM and CDMA networks are shown in Table 4–9 and Table 4–10.

Table 4–9 Typical GSM Network Data Transmission Speed

Standard	Data Transmission Speed
Short Messaging Service (SMS)	N/A
Circuit Switched Data (CSD)	9.6 Kbps
High-Speed CSD	Up to 57.6 Kbps
GPRS	Up to 171.2 Kbps
Enhanced Data GSM Environment (EDGE)	Up to 384 Kbps

Table 4–10 Typical CDMA Network Data Transmission Speed

Standard	Data Transmission Speed
SMS	N/A
CSD	14.4 Kbps
cdma2000 1xRTT	Up to 144 Kbps
Wideband Code Division Multiple Access (WCDMA)	384 Kbps to 10 Mbps

Data networks. In addition to cellular networks, wireless data networks also exist to provide wireless data services (see Figure 4–8). Two of the most important data networks are:

- Mobitex
- DataTAC

At present, the primary vendor for Mobitex is Cingular. Both Palm and RIM BlackBerry mobile devices use Mobitex in addition to several other networks.

Wireless data networks have a big advantage over cellular networks because the coverage is currently much greater. However, as Table 4–11 shows, the data transmission speed for both Mobitex and DataTAC is quite slow.

Bluetooth. Bluetooth technology allows users to wirelessly connect various computing, electronic, and telecommunications devices such as printers, cellular telephones, PDAs, Tablet PCs, and Laptop PCs.

Bluetooth is a radio frequency specification for short-range communications between devices and is meant to replace connection wires. As such, Bluetooth works well within a range of 33 feet (10 meters). Bluetooth uses radio technology that does not require line-of-sight between devices. As a result, it is more flexible and convenient than infrared. The

Figure 4–8 Data networks

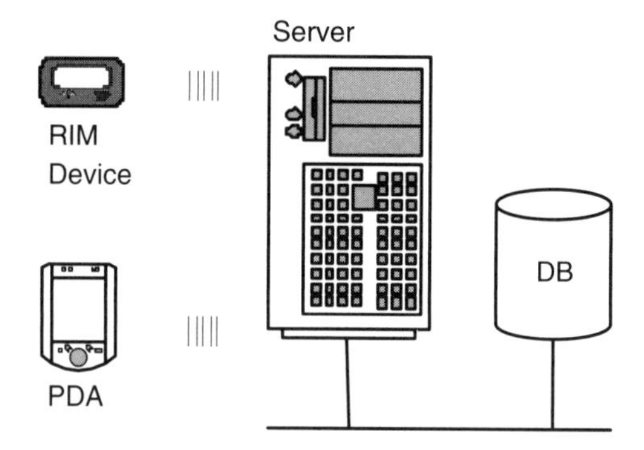

Table 4–11 Typical Wireless Data Network Data Transmission Speed

Standard	Data Transmission Speed
Mobitex	8 Kbps and 19.2 Kbps
DataTAC	19.2 Kbps

Table 4–12 Typical Bluetooth Data Transmission Speed

Standard	Data Transmission Speed
Bluetooth	56 Kbps–721 Kbps

gross data transmission rate for Bluetooth is 1 Mbps, although the actual rate is much lower, as shown in Table 4–12.

A Bluetooth-enabled mobile device allows you to synchronize your Pocket PC with a host computer. In addition, you can transfer files between Laptop PCs, exchange data with your cellular telephone, connect to the Internet, and send email from your Laptop PC via a cellular telephone without any cables. You can also print wirelessly and file, browse, and navigate on another Bluetooth device.

A user may form a Personal Area Network (PAN) containing multiple mobile devices that communicate with each other wirelessly using Bluetooth technology (see Figure 4–9). A collection of connected Bluetooth devices is called a "piconet." Up to eight devices can be in one piconet, and 10 piconets are allowed in a 10-meter radius. To establish a piconet, one of the devices must act as a master and the other devices as slaves.

Wireless LAN. A Wireless LAN is a local area network that uses wireless technology to communicate with mobile devices. Wireless LANs are often set up in corporate offices to allow users to roam throughout the building while maintaining a network connection. A Wireless LAN Card can be inserted into a mobile device (such as a Laptop PC), which allows wireless connectivity to a network.

Figure 4–9 Bluetooth Personal Area Network

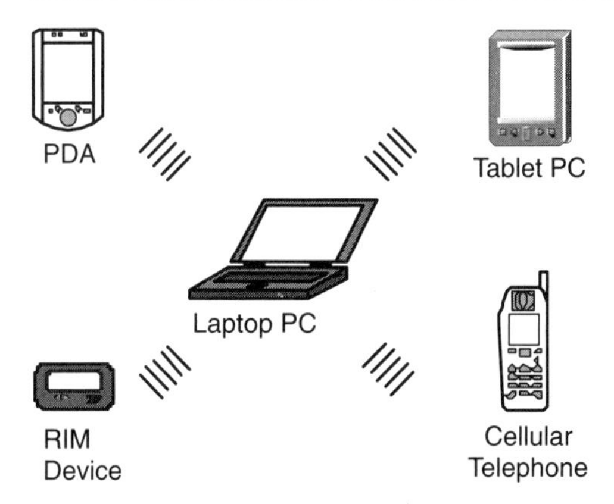

Table 4–13 Wireless LAN Data Transmission Speed

Standard	Data Transmission Speed
Wireless LAN 802.11b	1 Mbps and 11 Mbps
Wireless LAN 802.11a	54 Mbps
Wireless LAN 802.11g	54 Mbps and compatible with 802.11b

There are several Wireless LAN standards, including 802.11b. The 802.11b standard is also commonly known as Wireless Fidelity (WiFi). This is a high-speed line, as indicated in Table 4–13.

This technology has been extended to so-called "hotspots," which are places open to the public where wireless Internet access is provided. Typically, public hotspots include coffee houses, hotels, and airports throughout the United States. Verizon has also set up public telephone booths in New York City as wireless hotspots.

Satellite networks. Satellite networks provide another alternative for wirelessly connecting mobile devices.

One popular use of satellite networks is GPS. GPS is a satellite-based navigation system that utilizes satellites placed into orbit by the U.S. Department of Defense. These satellites orbit the earth while transmitting signals. A mobile device with a GPS receiver picks up these signals and uses them to triangulate the position of the device (see Figure 4–10).

GPS requires line-of-sight between a device and the satellites. In addition, the mobile device must receive a signal from at least three satellites to determine its latitude and longitude and will need to receive a signal from at least four satellites to determine altitude.

Figure 4–10 Satellite network

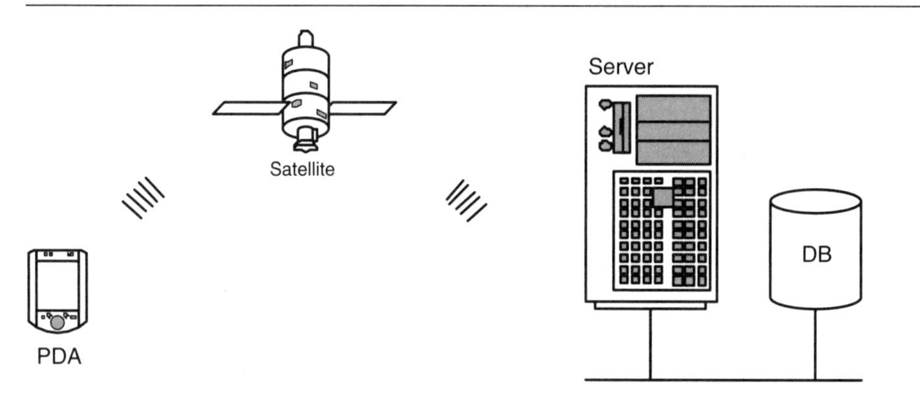

Today, there are many multi-functional mobile devices that contain GPS receivers. For example, Garmin makes a range of GPS-enabled products, including PDAs, cellular telephones, and walkie-talkies. In addition, many automobiles now include GPS-based dashboard navigation systems to assist drivers with location services.

4.4 SUMMARY

In this chapter, we discussed mobile device types and their components and features in some detail. We also discussed the connection methods used to connect mobile devices and back-end systems.

There are many mobile devices on the market and not all of them will suit all users, nor are they all capable of the same functions. What may suit one group of users may be unacceptable to another. Functionality, capability, portability, and cost are all considerations that must be taken into account.

CHAPTER 5

Mobile Client User Interface

We are just statistics, born to consume resources.
—Horace

A mobile client user interface assists the user with issuing instructions and receiving information from applications on the mobile device. A good user interface improves the overall user experience; a poorly designed user interface can jeopardize the mobile application's usage.

In this chapter, we initially consider the basic input and output (I/O) mechanisms that are used for communication between a user and a mobile device. We then discuss various types of application content and examine the user experience in terms of what a user looks for when using mobile applications. We finally consider a set of best practices for developing mobile client user interfaces.

5.1 USER INTERFACE

The mobile client user interface is literally a mobile device's interface with a user. As such, it is the primary area of interaction. Additionally, a mobile device may be connected to a network and a set of back-end systems and applications, and the user may thus also be indirectly interfacing to these systems and applications.

A user typically inputs information to a mobile device using one or several peripherals, such as a keyboard, mouse, stylus, or pen. The mobile device outputs information to the mobile device screen in the form of pages. These pages are ordinarily web pages or Windows Forms (see Figure 5–1).

In the following sections, we will consider the user interface characteristics of several mobile device types, including cellular telephones, PDAs, Tablet PCs, and Laptop PCs.

5.1.1 Device I/O

A user is able to communicate with a mobile device through a variety of methods. Typically, device input methods include a keyboard, mouse, stylus, pen, touch-screen, scanner, and voice commands, while device output methods include a screen, printer, and speech (see Figure 5–2).

Figure 5–1 User interface

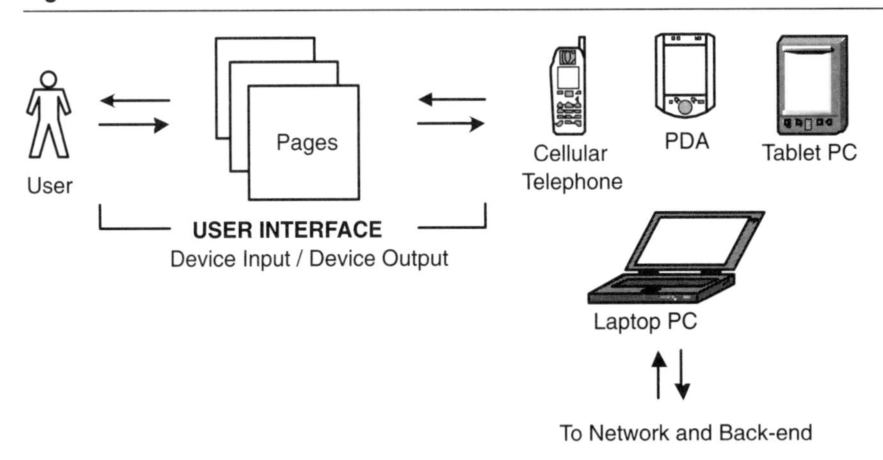

Figure 5–2 Device I/O

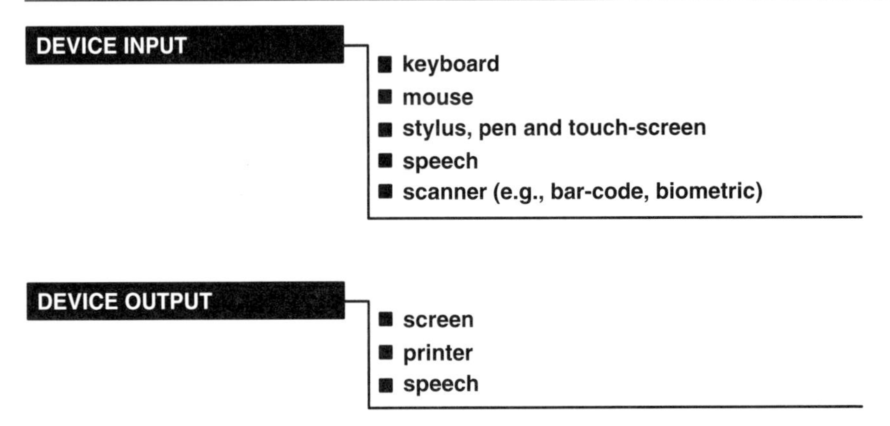

The mobile device peripherals that allow users to perform input and output may be small. For example, a cellular telephone has a tiny screen, while a PDA such as a Pocket PC also has a small screen and virtual keyboard.

These are challenges for the mobile application developer. For example, if the user is required to do a lot of reading, the challenge is to display information in a succinct and usable form on a small screen. Similarly, if the user is required to do a lot of typing or writing, the challenge is to allow the user to input information rapidly and accurately while reducing the amount of typing as much as possible.

In the following sections, we will discuss the various device input and output types in greater detail.

5.1.2 Keyboard

A keyboard's importance in the input of information varies considerably among mobile device types. For example, a keyboard is probably still the primary input device with mobile devices such as Laptop PCs. However, while a Tablet PC still has a keyboard, the intent is that the keyboard is relegated to a backup role with a stylus or pen supplanting it as the primary mode of input.

In contrast, PDAs generally do not have a keyboard at all but provide a virtual keyboard on the screen instead. This mode of entry has led to problems for users who need to type at any length because of the discomfort and inaccuracies involved in prolonged typing on small keyboards. In order to alleviate this, some ingenious mechanisms have been developed. For example, collapsible or folding keyboards allow you to enter large amounts of data quickly.

Cellular telephones also do not generally have a keyboard, although certain models with greater functionality do have collapsible keyboards.

5.1.3 Mouse

The traditional idea of an external mouse has changed somewhat in recent years, especially where a mobile device is concerned.

In the case of Laptop PCs, for example, a mouse is still used extensively, but it is now built into the device. However, certain users still use an external mouse and accept the inconvenience of carrying a little additional weight for the comfort and flexibility of the external mouse.

However, with mobile devices such as Tablet PCs and PDAs, a stylus or pen has largely—if not completely—replaced the mouse as the primary pointing and selection device.

5.1.4 Stylus, Pen, and Touch-Screen

The ability to write directly to a touch-screen using a stylus or pen is a powerful feature that is primarily identified with PDAs and Tablet PCs. In many ways, it obviates the need for a mouse and keyboard since a user is able to use a stylus or pen to select items as well as write and draw on the screen.

Unfortunately, prolonged writing on a small PDA by hand is difficult because the device is so small. It is also notoriously difficult for a user to enter information accurately and for the PDA to accurately recognize and interpret a user's handwriting.

In contrast, handwriting recognition on a Tablet PC is a much stronger feature and should prove to be an invaluable supplement to typing on a keyboard. The larger touch-screen also makes prolonged writing a more practical option (see Figure 5–3).

Figure 5–3 Tablet PC and stylus

5.1.5 Speech

Although recognized as an input method, speech is more regularly used as an output method (see Figure 5–4). For example, mobile devices can read aloud news, stories, and other information to users. Certain smart devices can also perform self-diagnosis and audibly report status or problems to users. Driving directions may also be given verbally, which may be a safer option than looking at a map on a screen while driving. Visually impaired users can be given verbal assistance. Alternatively, users may simply listen to stories for entertainment purposes.

In terms of voice command input, some advanced technologies exist, including Microsoft Agent. For example, a user can query a Microsoft Agent (e.g., "Peedy") for assistance through voice input. These agents use ActiveX technology and have a software development kit (SDK) that you can use to create your own caricatures and program your own agent.

Generally, however, the use of voice commands and speech recognition as user input is still in its infancy and the results so far are somewhat mixed. Voice commands are often not recognized by mobile devices and their vocabularies are still somewhat limited. Thus, it does not appear that speech will replace the other device I/O mechanisms as a major input method anytime soon, although speech does have many benefits from an output perspective.

5.1.6 Scanner

Scanners are periodically used as input devices. They are not generally intrinsically integrated into mobile devices because the types of input they take are somewhat specialized (see Figure 5–5).

Figure 5–4 Speech

Figure 5–5 Scanner input devices

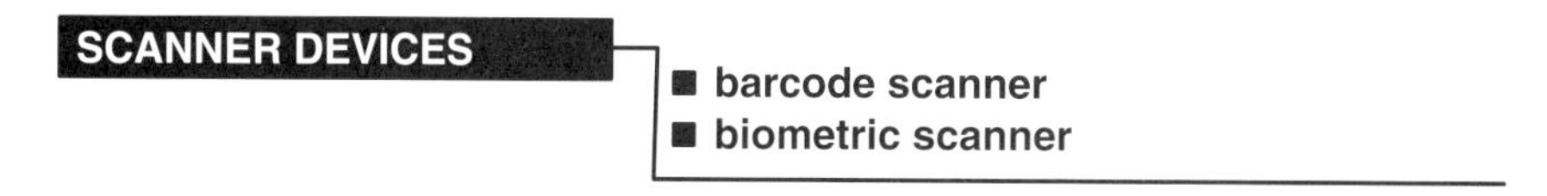

However, when scanning devices are used, they can greatly enhance a user's ability to input data. For example, barcode scanners are very useful in factory and warehouse environments where inventory typically has to be tagged very quickly, accurately, and efficiently.

The ability to uniquely identify a user through the use of a variety of biometric mechanisms such as facial recognition, fingerprinting, and iris or retinal recognition is an important feature of a user interface from a security perspective. Through such mechanisms, a mobile device is able to grant or deny access to a variety of applications based on the user's identity.

5.1.7 Screen

The specific characteristics of a mobile device screen (such as the display type and display size) affect the mobile application user interface. It is important to consider them when you are developing a mobile application (see Figure 5–6). These characteristics are described in more detail below.

Display type. A display has a variety of intrinsic characteristics. Some cellular telephones and RIM devices are monochromatic, while PDAs, Tablet PCs, and Laptop PCs have a color display. Therefore, if you are relying on color to present information (e.g., multicolored charts), it is also important that the values are distinguishable even if the display is monochromatic (e.g., by using cross-hatching patterns and different filler characteristics).

Display size. The display size of various mobile devices is of importance when user interfaces are considered. Other than altering or changing the mobile device type, it is generally not possible to increase the basic display size of a screen. In other words, if you are developing a mobile application for a cellular telephone, you may be limited to a very small display size that cannot be made larger.

Figure 5–6 Screen display characteristics

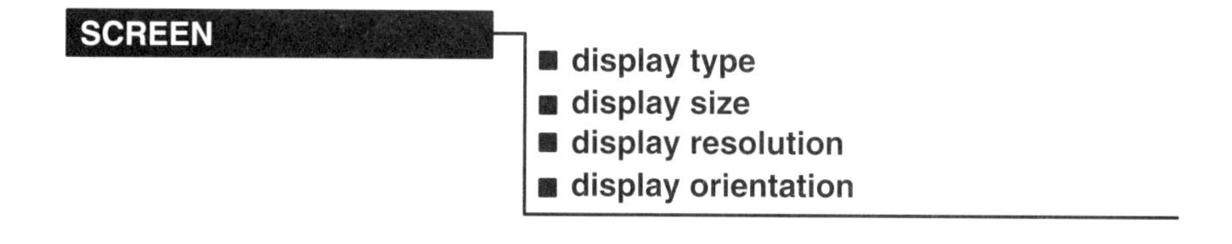

Figure 5–7 Display orientation

Display resolution. A screen typically has a maximum resolution, which is usually related to the device's quality and cost. A high-resolution screen is often very desirable but is also expensive.

Display orientation. In an attempt to utilize as much screen real estate as possible, some mobile devices—such as Pocket PCs—allow you to work in both vertical and horizontal mode (see Figure 5–7).

While apparently useful, this feature also means that a user has to learn a different way of looking at things. For example, buttons that are hard-wired at the bottom of the device are now on the right. At the very least, this would mean the normal order of the device's buttons is askew. In addition, the mobile device's ergonomic features may be subtly different. A handheld device designed so that an adult can comfortably grasp it may not be as comfortable to use when rotated 90 degrees.

Thus, while this is useful in some situations, horizontal orientation is not generally desirable since the bulk of Pocket PC applications are ordinarily designed to run vertically.

5.2 APPLICATION CONTENT

The mobile client user interface typically displays application content in the form of pages. Application page content can be rich or simple. *Simple* content typically consists of text only, while *rich* content includes text, images, animation, sounds, and movies (see Figure 5–8).

It is important to appreciate that not all mobile device types support all content types. For example, it is not possible or practical at present to watch movies on a cellular telephone or a PDA. On the other hand, it is perfectly possible to watch a movie on a Tablet PC or a Laptop PC.

The application content and layout constitute a mobile application's "look and feel." In the following sections, we will discuss the various types of content that you can provide to users and some of the tools you can use to generate and display that content.

Figure 5–8 Application content types

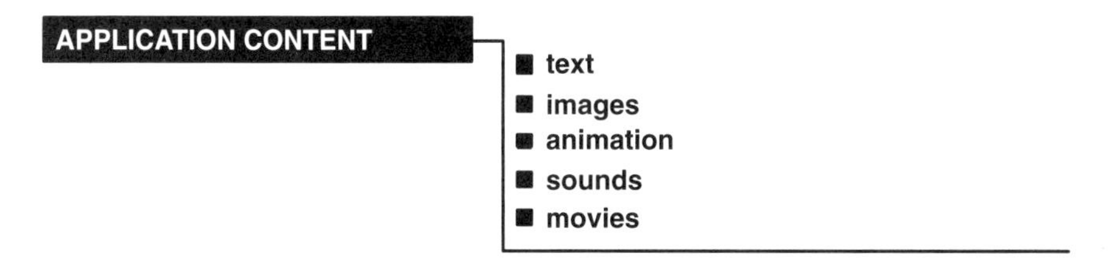

5.2.1 Text

Text is perhaps the single most important area in terms of application content. Good text content will hold the user's attention. Without it, a user will become bored and will not use your mobile application. It is also probably safe to say that text is the only universal medium for all mobile devices (the other media types may or may not be displayable).

Text can be created using any of a number of word processing editors, including Microsoft Word, Microsoft Notepad, and Adobe Acrobat. Formats include TXT, DOC, and PDF files.

Creating good text content follows the same rules as good writing composition. The text needs to have a purpose and provide the appropriate information for that purpose. In addition, it is not unreasonable to expect a mobile application's text content to be well-written, informative, and free of typographical errors, jargon, and slang.

If there is a lot of text, you should attempt to structure it within tables or forms. Tables can present huge amounts of data much more succinctly than paragraphs of text. Also, different font types and sizes, highlights (e.g., bolding, italics), and different text colors can be used to build effective and handsome mobile application pages with a low download overhead.

Ideally, you should allow text to wrap and resize automatically according to proportions of the screen. In addition, if there is a lot of text, don't make it too small. Also try not to use moving, scrolling, or blinking text unnecessarily since it can be distracting. Consider your audience. If they are visually impaired or older, ensure that the text is large and clear.

The information must also be timely and accurate. For example, if the application pertains to inventory management, information displayed must accurately reflect current inventory levels, or the user will lose confidence in the application. If the application is a news application, news must be provided in a timely fashion (see Figure 5–9).

5.2.2 Images

Images are probably second only to text in importance to application content. Images can be created and manipulated using any of a number of common software products, including Adobe Paint Shop Pro, Corel Draw, Corel Photo-Paint, Microsoft Paint, and Microsoft Visio. These

Figure 5–9 News text

NEWS
Home | World | US | Europe | Asia
Last updated: 10 minutes ago...
Headline 1
U.S. story...
Headline 2
Europe story...
Headline 3
Asia story...

products store images in a variety of file formats, including BMP, GIF, JPG, TIF, and MPG files. GIF and JPG are the most common graphics formats recognizable by modern web browsers.

Images can also be acquired through the use of digital cameras and image scanners. If you have a digital camera, you can take photographs of items, edit them, and prepare them for mobile application use. For example, some of the images in this book were generated by taking photographs and then using Microsoft Paint and Microsoft Visio to work with them (see Figure 5–10).

Figure 5–10 Image taken with digital camera

5.2.3 Animation

Animation is often used in advertising on mobile and web applications and to enliven an application.

Technically, animating image files is surprisingly easy to do, although the composition of animation sequence itself can be tricky. Essentially, you make a sequenced set of GIF images that contains the image, the image shifted a little, etc.—just like a cartoon flipbook (see Figure 5–11). This can be done using one of the imaging techniques described in the previous section. After the image sequence is complete, you make a single composite GIF file that contains the animation sequence. This can be done using a GIF animation tool which can be purchased as inexpensive shareware.

You can also animate buttons. For example, you can create three image files, the first containing a red circle, the second an orange circle, and the third a yellow circle. Once animated, the button would appear to change color from red to yellow.

When you animate a page, it is important to consider your audience. If the page they are reading contains a lot of text, you should use animation sparingly, since blinking, flashing, or rotating images can be extremely distracting. On the other hand, if you are trying to advertise something, you can get a little flashier to get your point across.

5.2.4 Sounds

The use of sounds can serve to improve a user's experience with a mobile application. Sounds may be recorded and converted into a format that is suitable for playing on a mobile device.

There are many mobile device audio players, including Microsoft Windows Media Player and Real Networks Real One Player. These players recognize a variety of audio file formats including Music CD Playback (CDA), Audio Interchange File Format (AIFF), Windows Media

Figure 5–11 Animation sequence of a computer virus attacking a firewall

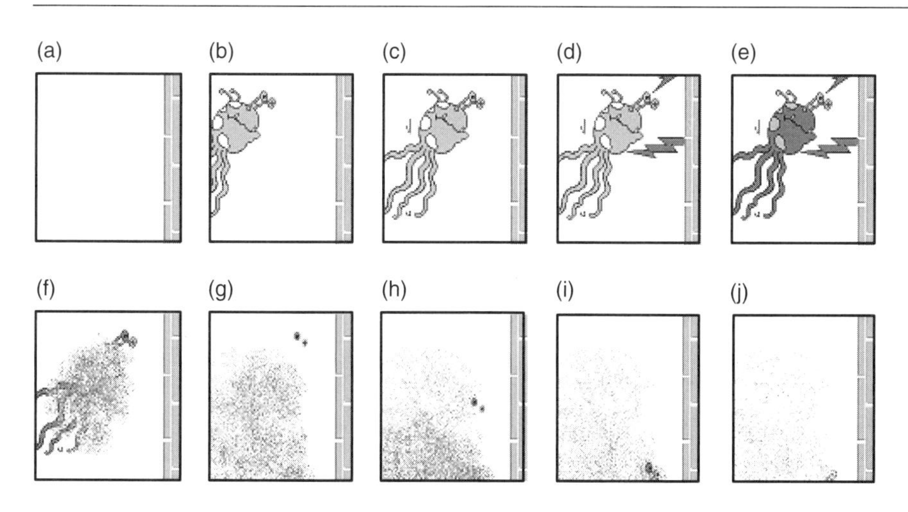

audio and video (WMA), Windows audio and video (AUI and WAV), Musical Instrument Digital Interface (MIDI), and MP3 files.

However, sound should be used judiciously. Most applications run relatively quietly, with an occasional beep or sound played when there is a problem or to highlight some event occurring on the mobile device. Generally, it is not a good idea to provide extraneous music or sounds unless that is the specific intent of the application. Applications that produce noise tend to become irritating after a while.

In addition, continuous beeping, noises, and other sounds can annoy not just the user but also those around the user. In fact, many users will mute the sound on their devices to prevent this problem, so you should not rely on sound alone to convey important information.

5.2.5 Movies

There are many tools available to help you capture, edit, create, transform, and deliver movie images in digital form. Once in digital form, there are several multimedia players that allow users to play this content on mobile devices.

For example, the Real Networks Real One Mobile Player allows you to play streaming media content on PDAs and cellular telephones. Microsoft Windows Media Player also recognizes and is able to play streaming video media, including Windows Media audio and video (WMA), Windows audio and video (AVI and WAV), Moving Picture Experts Group (MPEG), DVD video (VOB), and Macromedia Flash (SWF) files.

However, playing movies should be performed judiciously. The quality of the images is only marginally acceptable on most mobile devices at present. In addition, if you are accessing movies online, the device's communications bandwidth may not be sufficient to play them successfully. Furthermore, attempting to view movies on devices such as PDAs may also consume all of a device's memory and temporary storage.

5.3 USER EXPERIENCE

This section discusses a typical mobile client application from a user's perspective in terms of what the user sees and wants in order to work on the mobile application.

Generally, users want a highly interactive, content-rich application with pertinent text and images, which may be supplemented by animated images and other media. This content is typically laid out on a page.

If there is a substantial amount of application content, the application must have excellent navigation features along with help and search capabilities. Personalization is also important, allowing each individual the ability to have personal views of the application (see Figure 5–12).

The following sections will describe the above features in more detail. These features are based on our experiences with a wide range of users. However, when you actually carry out mobile application development, you should verify these generalities with your user population.

Figure 5–12 User experience features

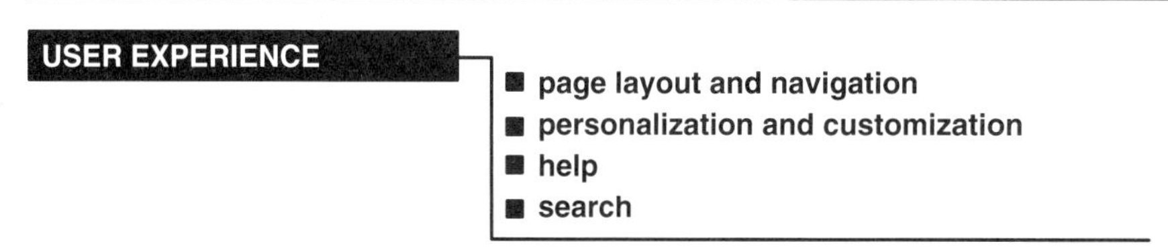

5.3.1 Page Layout and Navigation

A critical part of the user experience is the ability to easily navigate the application in order to view application content. In its most general form, application navigation is carried out by a set of menus, each of which contains a number of additional submenus until a final set of choices is presented.

Perhaps the single most important thing to remember when developing an application's navigation is that the user must never become lost in the application. The user must also be able to navigate to any part of the application using as few keystrokes as possible (often just one or two clicks). If the navigation is designed properly, even very large applications are easily navigable.

In the following sections, we will discuss several typical page layouts. These include a large web page and a large Windows Form (generally used for Tablet PCs and Laptop PCs), and a small web page and a small Windows Form (for Pocket PCs). (See Figure 5–13.)

Large web page layout. If you visit enough web sites, you will see that in many ways the web pages all follow a general pattern (see Figure 5–14). The top part of the page is typically taken up by a banner. This banner ordinarily contains a company logo and/or the application name. There might also be an advertisement in this area if the site is a commercial web site.

Just below the banner lies the main navigation area. This contains one to many menus that allow access to submenus. Each submenu item contains a set of options that may be viewed in the left-hand navigation bar.

On the right side of the page is the content area, which contains much of the page's actual content. Content may include text, images, animations, sounds, and movies.

Figure 5–13 Page layouts

PAGE LAYOUT

- large web page
- large windows form
- small web page
- small windows form

Figure 5–14 Large web page layout

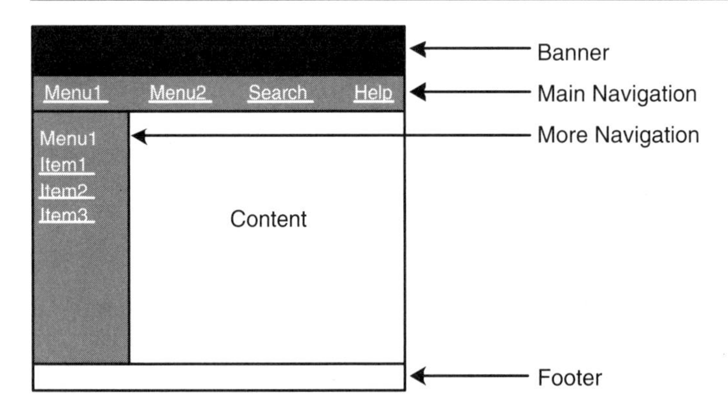

At the bottom of the page, there may be a footer that contains a copyright notice or links to supplementary areas of functionality.

While menus or lists of hypertext links can be placed anywhere on a web page, menus are typically placed horizontally at the top of a page or as a list on the left-hand side of a page. Upon hypertext link selection, some action is taken, usually resulting in the display of another page. In this fashion, users can navigate from page to page, where different actions can be taken.

Large Windows Form layout. In a Windows application, the top part of the form is typically taken up by a banner (see Figure 5–15). This banner ordinarily contains the application's name. The banner may also contain buttons for minimizing, maximizing, and closing the application screen.

Figure 5–15 Large Windows Form layout

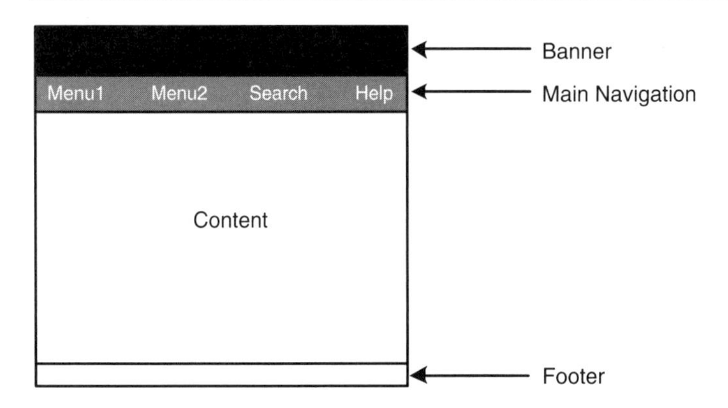

Just below the banner lies the main navigation area. Microsoft Windows menus are typically lists of choices that contain additional menus until a final list of choices is presented. The menus are usually placed at the top of a page and drop down to offer choices. Upon selection, a choice triggers a function that takes some action, typically resulting in the display of another page that can be viewed in the content area. Content may include text, images, animations, sounds, or movies. In this fashion, users can navigate from form to form where different actions can be taken.

At the bottom of the form is a footer that consists of supplementary and support information.

Small web page layout. In small web pages, the top part of the page is typically taken up by a banner (see Figure 5–16). This banner ordinarily contains a company logo and/or the application name. There might also be an advertisement in this area if the site is a commercial web site.

Just below the banner lies the main navigation area. This contains one to many menus that allow access to submenus. Each submenu item contains a set of options that may be viewed below the main menu. Because of the small size of this layout, there is usually no left-hand menu.

Below the menus lies the content area, which contains much of the page's actual content. Content may include text and images.

At the bottom of the page, there may be a footer that contains a copyright notice or links to supplementary areas of functionality.

While menus or lists of hypertext links can be placed anywhere on a web page, menus typically are placed horizontally at the top of a page. Upon hypertext link selection, some action is taken, usually resulting in the display of another page. In this fashion, users can navigate from page to page, where different actions can be taken.

Small Windows Form layout. Generally, Pocket PC applications utilize a Windows Form style layout (see Figure 5–17). Two significant differences, however, are their use of the bottom menu and their use of a virtual keyboard.

Figure 5–16 Small web page layout

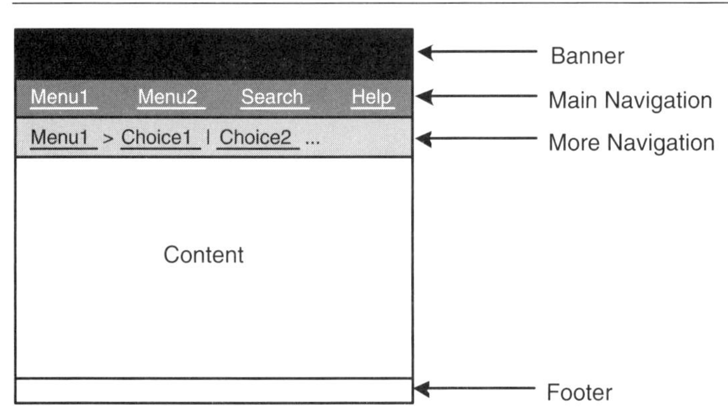

Figure 5–17 Small Windows Form layout

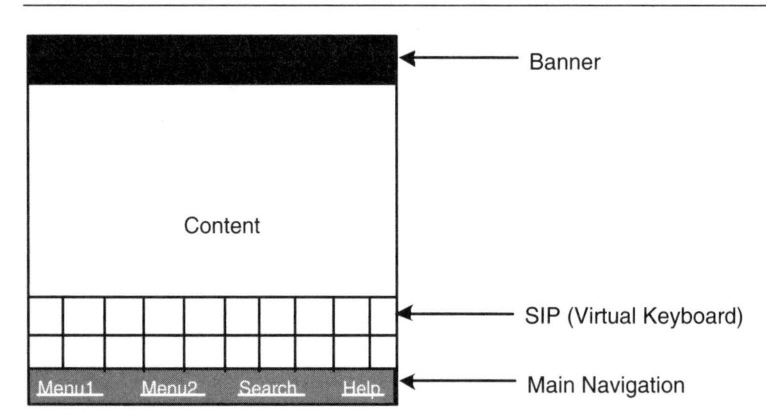

In a typical Pocket PC application, the top part of the form is taken up by a banner that ordinarily contains the application's name and a button for closing the application screen. It is important to note that on the Pocket PC this button hides the application but leaves it running, while on a Windows application this button actually stops the program.

Just below the banner lies the main content area. Content may include text, images, and input controls, such as text boxes and list boxes. When you are developing data entry screens, you should attempt to minimize the amount of typing needed. For example, you should replace a free-form text box with a combo box, list box, or radio buttons wherever possible.

At the bottom of the form lies the main navigation area. A Pocket PC form is somewhat different from a traditional Windows Form in that the menus are on the bottom of the screen.

Another difference between a Pocket PC form and a traditional Windows Form is that you may need to manage the display of the virtual keyboard or Soft Input Panel (SIP). Both Pocket PCs and Tablet PCs have a traditional SIP. When the SIP is displayed, it may block some of your other input controls. You will need to ensure that the SIP becomes visible when the user attempts to enter data into a text box and make the SIP disappear when it is not needed.

5.3.2 Personalization and Customization

Many web applications are written to allow users to personalize their experience. Personalization and customization of a web application can take several forms, including personalizing look and feel, personalizing access, and personalizing requested and directed content (see Figure 5–18).

Figure 5–18 Personalization methods

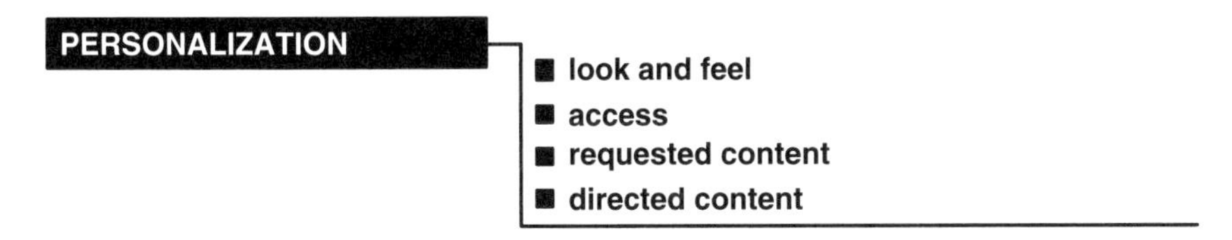

Personalized look and feel. Powerful web applications can provide users with the ability to change and personalize the look and feel of the application. Items that can often be personalized include:

- Page layout
- Navigation layout
- Content layout
- Content
- Background and foreground colors
- Text colors

Personalized access. Many web applications are written so that the user can use and see different aspects of the application. This depends on a variety of user attributes, including:

- User identity
- User group
- User entitlements

For example, an ordinary user might just be able to work in his/her application area. A user in a more privileged group might be able to work with more application areas or work with more application services.

Personalized requested content. A user may want to personalize the content provided by an application. Suppose you were interested in following a portfolio of stocks. On a simple non-customized web page you might click on a stock's link for the application to return the stock price (see Figure 5–19).

However, not every user is interested in the same set of stocks. What happens if a user wants his or her personal version of this web page? Users might also want stock prices sent directly to their mobile devices rather than going through a browser to obtain the information. Under these conditions, a simple web page becomes a little inflexible and cumbersome.

Figure 5–19 Non-personalized page vs. personalized page

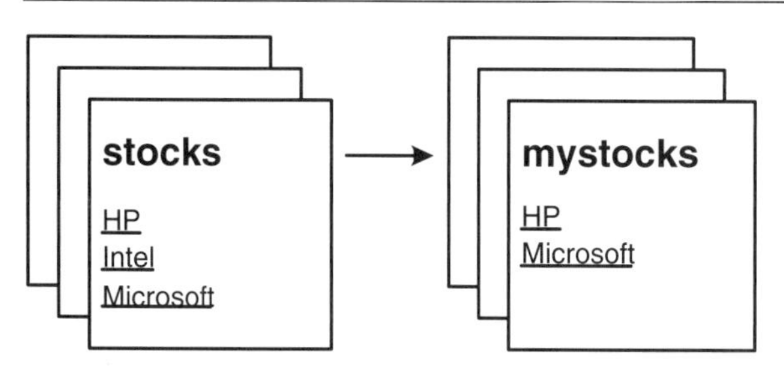

In order to personalize this page, a user makes certain personal information available to the information service provider, including his/her name, email address, and a list of stocks of interest. This information is stored on the server and allows the user to see his or her personal stock portfolio.

The demand for personalized content is very high, since it pertains to giving the user customized and specific information in a flexible fashion wherever they are.

Personalized directed content. The information that users provide can also be used to send users targeted content and advertising.

Using the above example, if an application service provider knows that you are interested in a particular stock, it could push information that it feels would be of interest to you. For example, it might email you news about your stocks of interest. It could also alert you or take some other action if a particular stock rose or fell.

5.3.3 Help

In all but the simplest of applications, a help feature is available (see Figure 5–20). Help can be presented to the user in a variety of ways, including:

General help. General help takes the form of textual and graphical information about an application. Other information that is often helpful to include within general help is a glossary of terms and a site map.

Context-sensitive help. Context-sensitive help is provided so that a user does not have to read through possibly large amounts of text in order to find the correct information.

Personalized help. If an application itself is personalized, the help text should also be personalized. For example, if a user is not able to see a certain section of a web application because he or she does not have the privilege to do so, any help text on that section also should be invisible. Personalized help is used for a variety of reasons, including security and applicability.

Figure 5–20 Help types

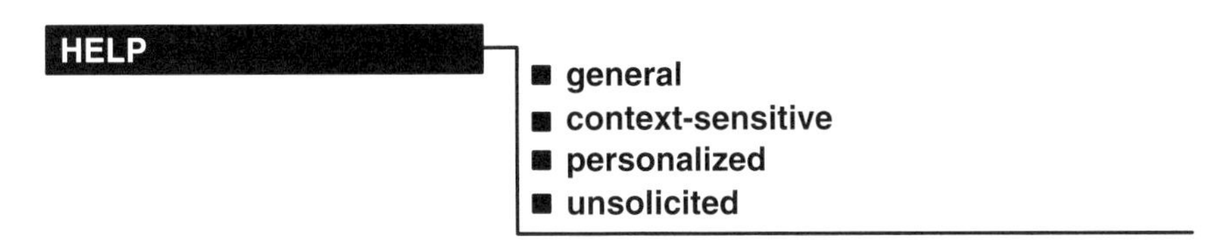

Unsolicited help. Some applications are able to determine what a user is doing (or attempting to do) and are able to offer advice and help accordingly. While this is a powerful feature, it should be used judiciously. Unsolicited help can annoy users, especially if the application has incorrectly determined what the user is trying to do. Experienced users may also find unsolicited help irritating, especially if there is no convenient way to turn it off.

5.3.4 Search

The ability for a user to search for information is often very handy. A commercially available search engine that assists users in finding text and image content is often implemented in large web applications. However, these search engines are not ordinarily available or supported on mobile devices.

5.4 BEST PRACTICES FOR DEVELOPING A USER INTERFACE

In the following sections, we will describe a set of specific practices that you might follow when developing user interfaces for mobile devices (see Figure 5–21).

Figure 5–21 Best practices

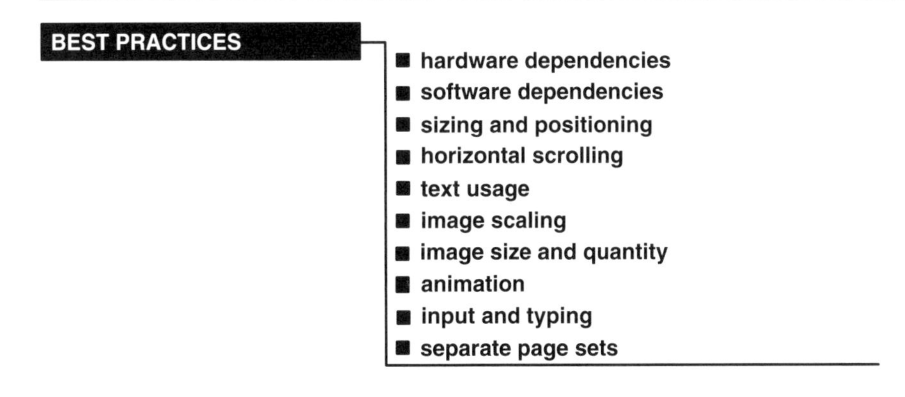

5.4.1 Consider Hardware Dependencies

Several considerations need to be taken into account from a hardware perspective when you develop a mobile application (see Figure 5–22).

The fundamental characteristics of each mobile device type affects the user interface. For example, it may be easy for a user to enter certain data on a Laptop PC, but it may not be possible to enter the same information using a PDA because the keyboard is too small. Similarly, other hardware dependencies previously discussed (e.g., mouse, screen, etc.) also need to be considered.

5.4.2 Consider Software Dependencies

As with hardware, several considerations need to be taken into account from a software perspective when you develop a user interface (see Figure 5–23). These are described below:

Browser type and version. There are numerous browser types and versions available. Cellular telephones, for example, typically use a WAP browser, while Pocket PC devices ordinarily use Microsoft Pocket Internet Explorer. Tablet PCs and Laptop PCs use Microsoft Internet Explorer, Netscape Navigator, Opera, and Mozilla, to name but a few.

Emulators and real browsers. While the use of emulators is very handy for development purposes, it is important to test your mobile application on a real browser prior to production deployment. Emulators and real browsers may not be the same, and code that may work on the emulator may not work on a real browser. When developing an application, testing should be done on all supported devices and browsers to avoid surprises when the application is rolled out into production.

Figure 5–22 Hardware dependencies

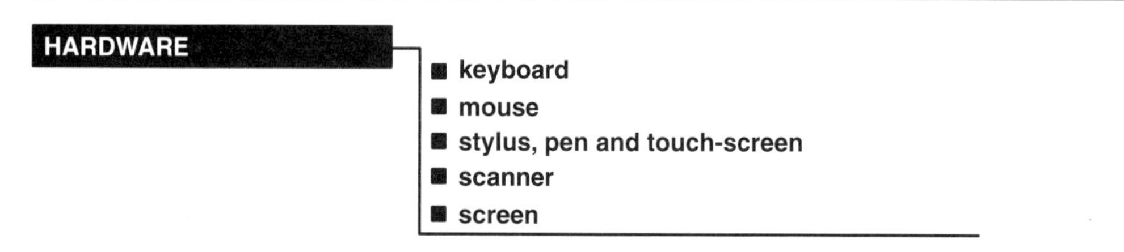

Figure 5–23 Software dependencies

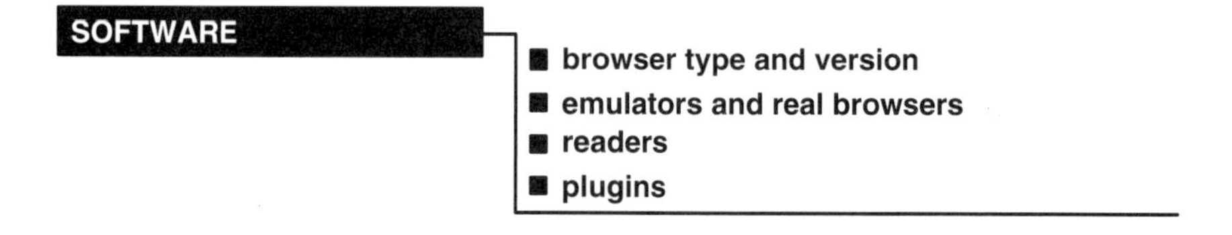

Readers. The required readers may or may not exist for the mobile device. For example, standard readers such as Adobe Acrobat Reader, Microsoft Word Viewer, or Microsoft PowerPoint viewer may or may not be available for the mobile device application you are developing.

Plugins. Similarly not all plugins are available for all devices. Plugins for Macromedia Flash and Microsoft Windows Media Player are available for Pocket PCs but may not be available for other devices.

Generally, there is a lack of readers and plugins, and not all document types can be handled on all mobile devices. In addition, mobile device readers and plugins may not function exactly like their larger counterparts. For example, Microsoft Pocket Excel does not support macros, a feature of Microsoft Excel.

It is reasonable to assume that, over time, more readers and plugins will become available for mobile devices. For example, CNetX Pocket Slideshow, which displays Microsoft PowerPoint files on Pocket PCs, and Adobe Acrobat Reader for Pocket PC are now available for use.

Nonetheless, there may be functions that will never be supported. Developers should check mobile devices' software dependencies since familiar products may not actually work in the same fashion to which you are accustomed.

5.4.3 Consider Absolute and Relative Sizing and Positioning

The ability to create rich content pages requires the ability to accurately size and position text and images on a page. You can specify sizes and positions through absolute and relative means. Each method has its pros and cons.

For example, suppose that you have a hypothetical mobile device screen that is 400 pixels wide and 400 pixels high. You want to place an image on this screen (see Figure 5–24). If you want this image to be exactly at *x*, *y* coordinates (99, 0), absolute positioning will allow you to do this. However, you cannot position the image precisely at (99, 0) using relative positioning. The closest you can specify is (25%, 0%), which means that the image will actually be located at (100, 0).

Figure 5–24 Absolute and relative positioning

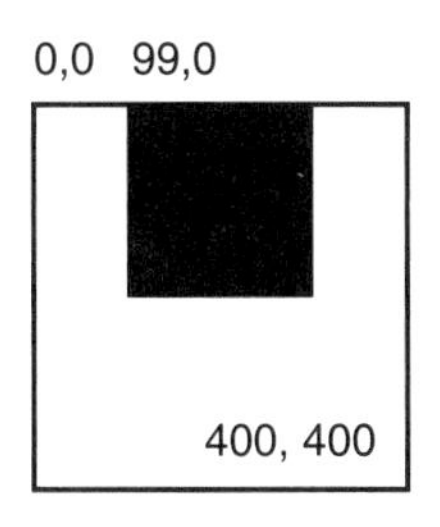

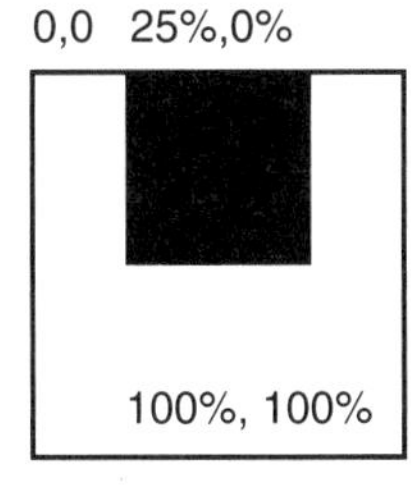

If you want to display the same image on a larger mobile device screen, 1,000 pixels wide and 1,000 pixels high, the image will still be located at (99, 0) as a result of absolute positioning. However, the image will automatically be located at (250, 0) with relative positioning.

Early web pages used relative positioning primarily with text and few images. When a table width was specified to 100%, the page would expand or contract the table depending on the size of the screen. For example, if the screen size was 15", the table would be set to 15". If the screen size was 10", the table size would be set to 10". Text contained within the table columns would automatically wrap accordingly and the tables self-adjusted according to the screen sizes. However, graphically, the control of pages and tables was not as fine and the look and feel was not considered to be of a high enough quality.

Later web pages used absolute positioning. This allowed developers to create very high-quality look and feel with very precisely placed tables, text, and images. However, this also reduced the ability for the same page to be displayed without modification when viewed on different device types.

Thus, a decision must be made: either exercise fine control through hard-coded dimensions (but have more display dependencies) or exercise less control using proportions (but have more display independence). It is also possible to mix absolute positioning with relative positioning.

5.4.4 Reduce Horizontal Scrolling

On mobile devices, you have limited screen space; it is important to make best use of this limited space. One key item from a usability perspective is to minimize or, preferably, prevent the need for horizontal scrolling.

A page of text that both horizontally and vertically scrolls makes the application all but unusable, since a user must horizontally scroll across every single line to read the text (see Figure 5–25).

In a mobile web application, the ideal is to set HTML element properties as relative proportions of the screen (e.g., set table width to 100%). Then, even when differently sized displays are encountered, a browser will automatically scale tables to fit and no re-coding has to be done.

Figure 5–25 Horizontal scrolling

Unfortunately, you can't exercise very fine control with proportions. In order to obtain this degree of control, you have to hard-code pixel dimensions into HTML pages. If an existing web site has a page designed to run on desktop computers with a table size of 600 pixels, the page cannot be viewed on a PDA without a horizontal scroll bar—the table size of 600 pixels far exceeds a PDA's screen width.

In order to reduce horizontal scrolling, a decision must be made: either exercise fine control through hard-coded dimensions (but have more display dependencies) or exercise less control using proportions (but have more display independence).

5.4.5 Use Text Extensively

A mobile device typically has limited resources and it is important to utilize them carefully. In addition, communication between a mobile device and a back-end application may be over slow communication links.

As a result, it is good practice to use text extensively, and reduce the use of large data files such as images, animation, sound, and movies. A well-written text-only mobile application (or a mobile application with very limited graphics) can still look remarkably elegant and work well because of lower communication overhead.

5.4.6 Check Image Scaling

The ability to successfully display images on various mobile devices depends on the size, quality, and scale of each image. An image may not necessarily be displayable on all mobile devices. One size does not necessarily fit all.

It may not be possible to scale up an image through the full range of mobile devices if the resolution of the image is not good enough on a larger scale. For example, it may not be possible to expand a passport-sized photograph of a person to a 4" × 6"-sized photograph without the image becoming grainy.

It may also not be possible to scale down an image through the full range of mobile devices if a specific subject in the picture is too insignificant. For example, a picture of a small cottage set in a large landscape of fields and sky may drop to insignificance when the image is reduced.

Thus, it may be necessary to perform some manual editing of images to maintain a high quality level. Automatic scaling may be possible, but it cannot be assumed that this will work.

In addition, large images take up a great deal of space. A 10-MB photograph is not uncommon. Automatic scaling of the image does not change the actual amount of space the image takes up. Therefore, if you are trying to transmit this photograph to a PDA over a slow line, it may take a long time. Even though you will only see a small image on the PDA, it may take a long time for it to arrive.

Relying on a browser such as Microsoft Pocket Internet Explorer to automatically scale images may not work, since images may not be clear after scaling down. In many applications, you may actually need to edit images and buttons so that they will fit onto a small screen without horizontal scrolling.

5.4.7 Reduce Image Size and Quantity

A mobile device typically has limited resources (e.g., CPU, memory) and it is important to utilize them carefully. In addition, communication between a mobile device and a back-end application may be over a slow line.

As a result, it is good practice to reduce image size and quantity. Try to make images as small and as narrow as possible. This has the benefit of reducing horizontal scrolling and also reducing download time. Also, reduce the image quantity if possible. Consider using only a few small images per page.

Consider users on slow data lines and users with limited memory and disk space (remember, browsers typically cache/store downloaded images on the user's device, taking up client disk space). Although many users at home have high-speed Internet connections, this is not an option on all mobile devices. For example, many cell phones only transfer data at 9.6–14.4 Kbps. Thus, any reduction in image size and quantity would be of enormous benefit.

5.4.8 Reduce Animation

If you intend to use animated GIF files in your mobile application, you should attempt to keep the size of these GIF files as small as possible.

Other related considerations include checking to see how many times you are looping through the animation sequence. If the looping is infinite, is it really necessary? And how rapidly are you looping?

5.4.9 Reduce Input and Typing

If the mobile application you are developing requires a good amount of data entry, it is important to remember that a user might have to enter the data through a stylus on a virtual keyboard. This can be both difficult and error-prone for the user.

Any reduction in the amount of actual typing the user has to do through the use of components such as drop-down boxes, radio buttons, and checkboxes is very helpful and will make the user experience more pleasant.

For example, in the case of the form in Figure 5–26, the use of the Group and Subgroup drop-down boxes and the radio button means that the user does not have to type "Group1," "G1 Administration," or "Yes."

In addition, you may be able to customize the virtual keyboard to make data entry easier in your application. This can be done by developing a custom SIP that contains letters, numbers, words, and phrases that the user frequently needs to enter into your application.

Figure 5–26 Reducing input and typing

There are, of course, limits to what can be done. If the user genuinely has to type a great deal on a Pocket PC on a regular basis, it may behoove the user to obtain a collapsible keyboard or use a device with a larger keyboard, such as a Tablet PC.

It is also interesting to note that the Microsoft Pocket Internet Explorer address bar is actually quite short (approximately 30 characters). A user who is browsing a web site that contains many pages with long URL names can have a difficult task determining which page is the correct one to select, since the addresses cannot be seen until the user clicks on the address and navigates to the end of the entry to find out (see Figure 5–26). Thus, keeping the URL short helps to reduce the amount of typing and difficult navigation.

5.4.10 Implement Separate Page Sets

If you are developing a mobile application that supports multiple mobile device types, it may be helpful to consider using separate sets of pages that are individually suited to specific mobile device types.

For example, if you need to develop a mobile application that works on both Pocket PCs and Laptop PCs, you may find that the look and feel and functionality of the displays are so disparate that you have to implement a separate set of web pages for each of them.

Many news sites typically offer this type of scenario (e.g., the British Broadcasting Corporation [BBC] web site). A full-featured, content-rich (text, images, sound, news clips, etc.) set of web pages can be viewed using a Tablet PC or Laptop PC. A text-oriented web site is also available for PDAs, while an even smaller version is available that pushes news headlines to cellular telephones and RIM devices.

The generation of these pages may be performed through the use of a "renderer." This is a program in the mobile application presentation layer that takes data from a data source and renders or writes it out in several different formats depending on the specific mobile device type. We will discuss renderers in more detail later.

5.5 SUMMARY

The mobile client user interface is a critically important area that developers need to consider when developing a mobile application. The greatest mobile application in the world technically will not succeed if the user does not find it easy, convenient, and helpful to use.

A mobile client application's content can include text, images, animation, sounds, or movies. However, only text and images can generally be relied upon as always or nearly always being displayable. Other media, such as sound, may or may not be supported by various browsers and displays. In addition, other media may be constrained by mobile device and communication characteristics, such as CPU or communication speed.

A great user interface attracts people to your application, while great application functionality and content holds them. The judicious use of text and images to create intuitive application navigation and personalized mobile applications also helps improve the user experience.

CHAPTER 6

Mobile Client Applications

Our doubts are traitors and make us lose the good we oft might win by fearing to attempt.

—Shakespeare

This chapter describes mobile client applications. It initially presents three mobile client application architectures and discusses the major architectural components of each of them. It goes on to discuss the pros and cons of each mobile client application architecture and some of the problems you may encounter when working with them. Finally, it discusses some of the best practices behind implementing these mobile client application architectures.

6.1 THIN CLIENT

Earlier in this book, we discussed the idea that mobile client devices can operate in three connectivity modes with back-end servers: always connected, partially connected, and never connected. We also discussed the fact that some mobile devices, such as RIM devices, are always on and always connected, while other mobile devices, such as PDAs, may only connect to a network intermittently.

When developing a mobile application, you must determine who will be using it, where they will be using it, and on what specific device. If the users have network connectivity whenever they need to use your application, you may find the use of a thin client to be ideal.

For example, you may be developing an application that allows doctors to enter patient data into a back-end system via a mobile device. If those doctors are working on hospital wards that have a Wireless LAN, then you may easily be able to develop a thin client solution. However, if the doctors will also use the application while they are out on house calls, you need to determine whether they will have access to the network during that time. If so, then you may still use a thin client application. If not, you will have to use one of the other means discussed in this chapter.

6.1.1 Development Considerations

Thin client applications are generally much less device- and operating system-specific than fat clients. This is because no custom application code is installed on the mobile device. Instead, users access server applications through industry-standard software, such as a web or WAP browser.

However, even though thin clients are less device- and operating system-specific, there are some device dependencies to consider, such as the size of the display area and the markup languages the browsers can interpret. Thus, you still have to decide which specific mobile devices you need to support and render the appropriate page content to those devices accordingly.

Pages may be generated through the use of a "renderer." This is a program in the server's Presentation Layer that takes data from a data source and renders or writes it out in several different formats depending on the specific mobile device type (see Figure 6–1).

In order to use a renderer, application data should be placed in XML files and a separate XSL style sheet for each supported mobile device should be developed. At run time, a renderer checks the user's browser type, picks the appropriate style sheet, transforms the XML, and renders content to the mobile device.

However, this approach can be unwieldy if you are supporting a large number of mobile device types, since you will need to develop multiple XSL files for each web page you wish to display. In order to help alleviate this, Microsoft supports multiple mobile device types through the use of the ASP.NET mobile controls and Mobile Internet Controls Runtime.

Figure 6–1 Renderer creating separate page sets

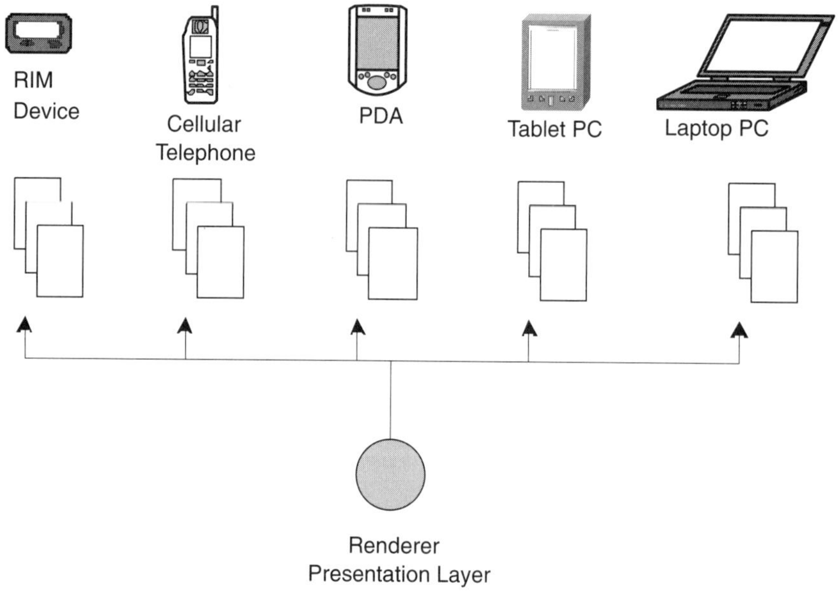

The runtime provides a rendering engine that translates ASP.NET web page code into different markup languages—such as HTML or WML—depending on the capabilities of the mobile device on which the content is to be displayed. The runtime contains device adapters that render the application on a variety of different devices.

In order to use this rendering engine, you must develop a set of web pages that use the ASP.NET mobile controls. These controls are a subset of the full ASP.NET Web Form controls.

6.1.2 Architecture

A thin client has no custom application code on it; all code resides on the server. The only software needed to run a thin client application is typically a web or WAP browser. The web browser may support some client-side scripting language such as JavaScript or VBScript for form validation. The server typically has a rendering engine as part of its Presentation Layer and is capable of displaying the appropriate content for that mobile device (see Figure 6–2).

Figure 6–2 Thin client application architecture

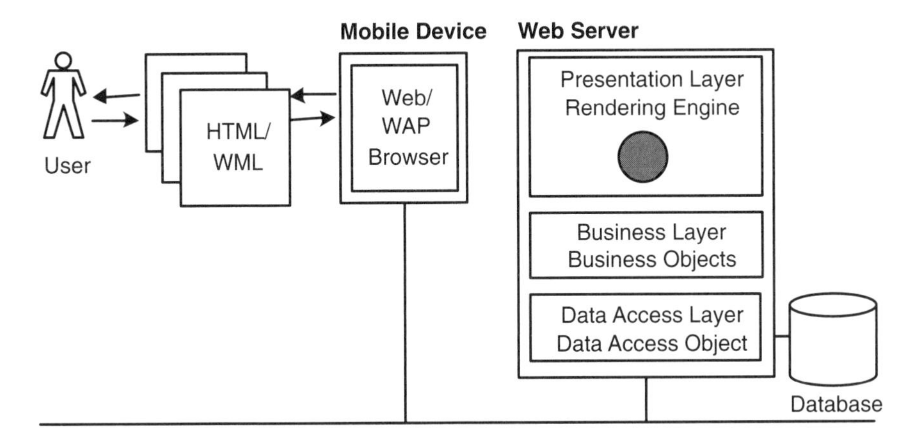

6.2 FAT CLIENT

Fat client applications have one or more layers of custom application code on the mobile client. They are often used when a mobile device is ordinarily only partially connected to the network. They may also be appropriate when you want to utilize a specific peripheral of a device, such as a camera, telephone, or GPS software.

For example, a doctor who makes house calls may download his daily caseload to his PDA while in the office. The doctor may then visit patients, where connectivity is not guaranteed. Nonetheless, the doctor may enter patient notes and store the data locally on the PDA using the fat client application. Then, when he returns to the office later, he may reconnect his PDA to the network and upload his notes to the back-end server.

6.2.1 Development Considerations

Fat client applications are generally device- and operating system-specific. For example, an application built for a Pocket PC will typically not work on a RIM device or Tablet PC. In this section, we will focus on fat client applications for Pocket PCs and Tablet PCs. Table 6–1 describes some options available for developing a fat client application using Microsoft technologies, such as Visual Studio.NET.

One advantage to using Microsoft Visual Studio.NET is that you can use the same development tools and languages to build your Windows, web, and mobile applications. It is important to bear in mind, however, that it may be difficult to reuse objects specifically built for one platform or another. For example, a Tablet PC runs the full .NET Framework, while a Pocket PC runs the .NET Compact Framework. Many of the .NET system classes have limited functionality on the .NET Compact Framework and developers may occasionally find it frustrating when methods they would like to use are not supported. The .NET Framework documentation in the Microsoft Developer Network (MSDN) library indicates the methods supported by the .NET Compact Framework for each class.

6.2.2 Common Use Cases

In this section, we describe some of the common use cases that a fat client mobile application must typically support.

TABLE 6–1 Fat Client Application Development Options

Device	Programming Language
Pocket PC	VB.NET (.NET Compact Framework) C# (.NET Compact Framework) eMbedded Visual C++ 4.0
Tablet PC	VB.NET (.NET Framework) C# (.NET Framework)

A use case is a short description of an activity or process that your application will provide. Use cases can be described using the Unified Modeling Language (UML) and illustrated as use case diagrams.

Each use case involves one or more *actors*. Actors can be people or systems. The use case describes the ways in which the various actors will use the application to complete a business process or function.

The following are several typical use cases that are common to many industries, including healthcare, government, finance, and the sciences (see Figure 6–3). These functions may be implemented in a similar way, regardless of the device or language you use.

1. **Log in.** Most mobile applications have some type of login mechanism. This is an important security consideration for mobile applications, since a device may be easily lost or stolen. While certain advanced devices have biometric readers to keep out intruders, most applications will still use some type of login page that prompts the user for a username, password, and/or other identifying information.

   The user's login credentials may be checked against the remote server or the local database, depending on the state of the device's network connection. If the user has successfully logged in remotely, you may save the login credentials in the local database. This will allow the user to log in locally when he/she is disconnected from the network.

   You may also want to allow a user to log in locally, even if the device is connected to the network. This will reduce lengthy round trips to the server if the user only wants to edit local data. However, if the user wants to download data from the server, or upload changes made locally, you should verify their credentials against the back-end server database.

2. **Select a task.** Once a user has logged in to the application, you will probably want to display some sort of form or menu that allows the user to select a task to perform.

   This task selection menu should be context-sensitive. For example, you should not show the user an option to download data if the device is not connected to the network. You also should not show the user an option to upload data, even if the device is connected to the network, if the user has not actually made any local changes.

   By displaying a context-sensitive task selection menu, you will prevent users from becoming frustrated by trying to perform inappropriate tasks.

3. **Download data.** When the user is connected to the network, you may allow him/her to download data to the device.

   To reduce the download time and storage necessary on the local device, you can only allow the user to download certain items from the server. For example, your application might only allow a doctor to download the patients who make up his/her caseload for that day instead of all his/her patients. The application could then download the full patient records for each patient selected.

Figure 6–3 Common use cases

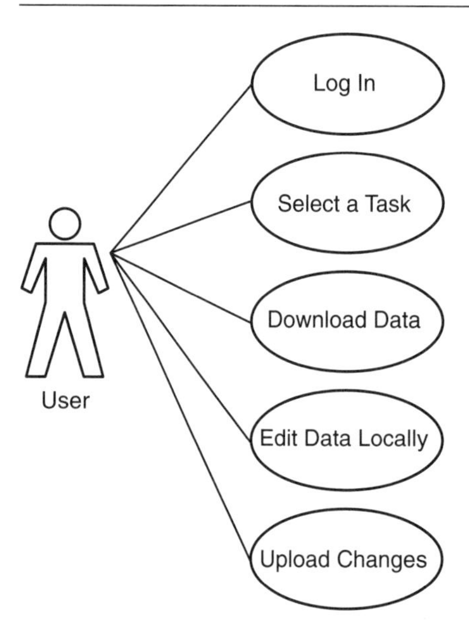

4. **Edit data locally.** Your application may also give a user the ability to search, add, and edit local data. These specific requirements will vary greatly depending on the requirements of your application.

5. **Upload changes.** When a user reconnects to the network, you may allow him/her to upload data. You may upload all data as a bulk load or allow the user to select certain items to upload to the server.

6.2.3 Architecture

This section describes the typical architecture of a fat client mobile application for a Pocket PC or Tablet PC. Although they use different technologies, there are common architectural guidelines that both devices should follow.

We will discuss the architecture of a fat client application with three layers: the Presentation, Business, and Data Access Layers (see Figure 6–4). The fat client application contains the following items:

- Windows Forms
- Business objects
- Data Access object
- Database

Each of these items is discussed in detail in the following sections.

Figure 6–4 Fat client mobile application architecture

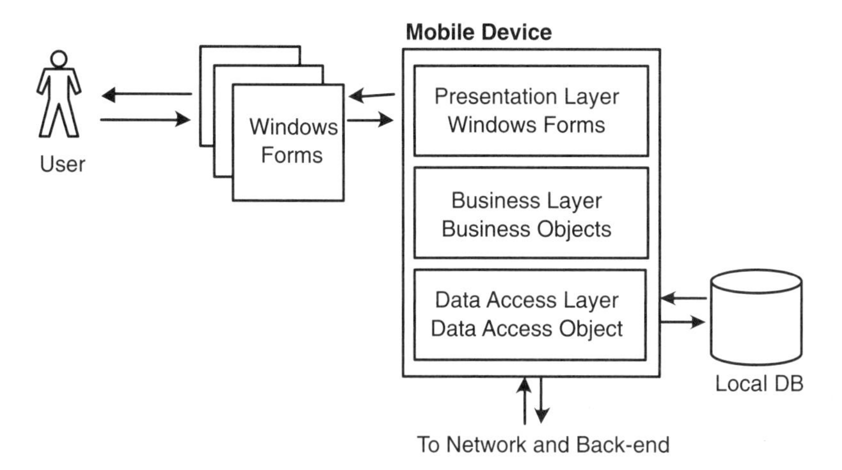

6.2.4 Presentation Layer

The Presentation Layer consists of Windows Forms and will use traditional Windows Form controls such as text boxes, list boxes, combo boxes, and radio buttons to display and gather data from the user. Look and feel considerations for the Presentation Layer were discussed at some length in the previous chapter.

6.2.5 Business Layer

The Business Layer consists of a set of objects that contain business and application logic. Some of the considerations when developing business objects include:

> **Object reuse.** You may have developed business objects as part of your server application and wish to reuse them in your mobile application. If so, you could try to deploy these same business objects to the mobile device. However, that may not be possible since you are not targeting the same platform.
>
> Another option is to expose your server-side business objects by creating an XML Web Service. This web service could reside on your server and forward method calls to your existing business objects. This method allows you to releverage some of your server code. Your mobile application can then communicate with your existing business objects whenever the device is connected to the network.
>
> However, you will probably also need to develop a second set of business objects to run locally when the user is disconnected from the network. These will probably be scaled-down versions of the server business objects since mobile applications typically do not provide the complete functionality of server applications. However, they should still contain the business logic that is critical to your original application.

For example, suppose that you have a business object running on your server that performs some validation on the data entered—such as checking that a patient's blood pressure is a realistic number—before allowing you to add patient medical data to the database. You cannot leave this business logic out of the mobile application. If you did, then a user might enter invalid data in the local database and only notice the error when the data was being uploaded to the server. By then it might be too late to test the patient again.

Object design. Your business objects may be designed using the tenets of object-oriented design. In object-oriented design, each object represents an entity in a real-world application scenario, such as a User or a Patient. Each object may have a set of properties and a set of methods. Properties represent the characteristics of an object. For example, a Patient object may have properties like Name, Age, Weight, and Blood Pressure. Methods are the functions and procedures that operate on an object. A Patient Object may have methods like "GetMedicationList" or "GetAllergyList." Some objects, such as those used for data communication, may not have real-world counterparts. We refer to these as *utility objects*.

Objects are often represented using UML class diagrams. Each class diagram shows the class name and its public properties and methods.

Figure 6–5 illustrates a class diagram for a Patient object. It contains the following elements:

1. The name of the class. This class is called "Patient."
2. The public properties of the class. This class contains the properties FirstName, LastName, etc.
3. The public methods of the class. This class implements the methods New, Save, Delete, etc. We will discuss these methods in the next section.

Figure 6–5 UML class diagram

Patient
-FirstName -LastName -DateOfBirth -Height -Weight -BloodPressure
+New() +New(in PatientId : Integer) +Save() +Delete() +GetPatientList() +GetMedicationList() +GetAllergyList()

Common object properties and methods. Generally, most objects will have some public properties. Some may be read-only while others may be modifiable. A property may be any data type, such as an integer, string, or date. In addition, a property may return a collection of child objects. For example, a Patient object might contain a collection of Allergies.

Most objects will also contain some of these common methods:

- **Constructor (also called "New").** The constructor is called when the object is created. The constructor can be overloaded to accept different parameters. For example, you might have one constructor that contains no parameters and only creates an empty object. Another constructor might accept a unique object ID as a parameter and use that ID to query the data source and populate all of the object's properties.
- **Destructor.** A destructor is called when an object is destroyed; it typically contains clean-up code.
- **Save.** Many objects will contain some method to save the state of the object to the underlying data store. The save method may perform an insert or an update depending on the business logic rules.
- **Delete.** Some objects may contain a method to delete the object's data from the underlying data store.
- **Get a collection of objects.** Some objects have a method that returns a collection of objects of that type. For example, a Patient object may have a GetPatientList method that queries the database for all patients, retrieves each patient's data, and returns a collection of populated Patient objects.
- **Get back-end data.** Many business objects for the mobile application will probably have at least one method that communicates with the remote back-end server. For example, one such method may retrieve a list of objects from the remote server represented as XML. Another method may upload changes made locally to the remote server. The details will depend on what type of data communication mechanism you are using between the device and server, such as SOAP, message queues, TCP/IP, etc.
- **Other business logic.** Objects may also contain many other methods used for business logic.

6.2.6 Data Access Layer

The Data Access Layer consists of the data access object and the database itself. Some of the considerations when developing the Data Access Layer include:

Data access object. Fat client applications may use a data access object to communicate with the local data store. Ideally, all communication with the database is carried out through this object.

The data access object provides a set of functions for manipulating the local database and performing queries. The code in this module will vary depending on the type of local database used.

The data access object will probably contain the following methods:

- **Create database.** This method creates a database with the connection string passed as a parameter. This connection string includes the database name and password. When developing an application for a Tablet PC, you will probably create the database as part of the installation package. However, in Pocket PC applications, you may need to create the database programmatically.
- **Check that database exists.** This method checks to see if the specified database exists. This is useful in Pocket PC applications, where the database is created programmatically.
- **Delete database.** This method deletes the specified database.
- **Execute SQL Command.** This method connects to the specified database and executes the given SQL statement.
- **Execute dataset.** This method executes the given SQL statement and returns a dataset.
- **Execute scalar.** This method executes the given SQL statement and returns the scalar value returned by the SQL call.

Local database. There are several options for storing data in a local database, including:

- **Microsoft Access.** This may be used on the Pocket PC or Tablet PC.
- **Microsoft SQL Server.** This may be used on the Tablet PC.
- **Microsoft SQL Server CE.** This version of SQL Server runs on the Pocket PC. SQL Server CE 2.0 is integrated with the .NET Compact Framework.
- **Other databases.** Other database vendors such as Oracle and Sybase also provide world-class databases that can run in the enterprise and on mobile devices.

Database synchronization. To this point we have discussed the client application code that reads and writes to the local database. In order for the store-and-forward mechanism to be complete, the locally stored data must eventually be forwarded to the server and synchronized with the main database. We will discuss data synchronization methods in the next chapter.

6.3 WEB PAGE HOSTING

An alternative to developing traditional thin or fat client applications is to host web pages on the mobile client. In order to do so, you need a web server that can run on a mobile device and is able to service web pages and business objects.

There are several benefits to doing this. For example, a user can still use the mobile application even when the mobile device is not connected to the network. In addition, you might be able to reuse existing server side web pages and business objects on a mobile device. Even if you cannot reuse your existing pages, it may still be easier to develop a second set of small web pages than to create a new fat client Windows Forms application.

In the following sections, we describe how this option works and what technologies are currently available to perform web page hosting on a Pocket PC.

6.3.1 Development Considerations

The idea of hosting web pages on mobile devices is somewhat unusual and has not been very well explored.

In order to service web pages on a client, you need the equivalent of a web server on the mobile client. Microsoft does include an HTTP Server with the Pocket PC 2002 and Windows Mobile 2003 SDKs. This HTTP server hosts HTML pages and simple ASP pages.

However, the current Microsoft Pocket PC HTTP Server does not support ASP.NET. This is problematic because Microsoft is discouraging application development for Windows Mobile 2003 that is not based on .NET.

Until Microsoft or another vendor releases a web server on a Pocket PC that can support .NET, we have created our own web server with partial .NET support called the Pocket Web Host (PWH). The complete PWH source code is available for download from this book's companion web site.

6.3.2 Architecture

This section describes the architecture of the web page hosting solution (see Figure 6–6). This discussion is based on the HTTP Server provided by Microsoft for the Pocket PC.

Figure 6–6 Web page hosting application architecture

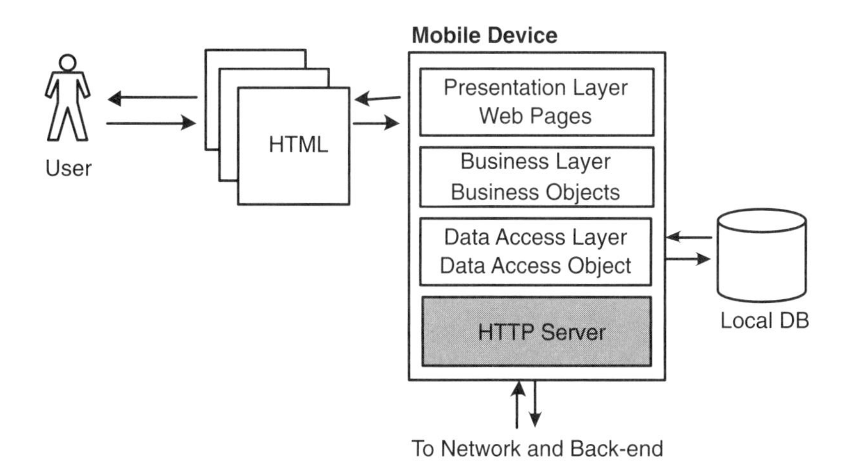

6.3.3 Presentation Layer

The Presentation Layer of the web page hosting client consists of HTML and ASP pages. These pages can be serviced on the mobile device even when the mobile device is not connected to a network and are viewable on a Pocket PC using Pocket Internet Explorer.

6.3.4 Business Layer

Ideally, the Business Layer of the web page hosting client contains a set of business objects. In reality, however, you will probably not create business objects for use with the Microsoft Pocket PC HTTP Server, since this would require creation of COM objects using eMbedded Visual C++. In practice, this may be quite difficult to do so you may end up keeping most of the business logic in your ASP pages, which is a less than optimal solution.

6.3.5 Data Access Layer

The current version of the Microsoft Pocket PC HTTP Server supports ASP pages that read and write data to a local SQL Server CE database using ActiveX Data Objects for CE (ADOCE). This local data will eventually need to be forwarded to the server and synchronized with the main database. The data could be synchronized from within an ASP page, although it might be preferable to develop a simple fat client application that does the synchronization since this could use more advanced technologies. We will discuss data synchronization methods in the next chapter.

6.4 BEST PRACTICES

The following sections describe some of the best practices for designing and developing mobile clients.

6.4.1 Consider Thin Versus Fat Versus Web Page Hosting

Many of today's mobile client applications are written using fat client technologies. While there are instances where this is necessary, this approach also has inherent drawbacks. For example, the fat client application is often device specific. Any changes to the fat client application can therefore be difficult, requiring the redevelopment of Windows Forms.

Release management can also be difficult with a fat client application since changes may require a rebuild of the application, which can affect hundreds of devices. In addition, a separate code base must be maintained for each mobile device type since the same files usually cannot be deployed to both a Pocket PC and a Tablet PC.

An alternative method is to employ a thin client implementation. However, this is only useful if the mobile device has a constant, uninterrupted network connection.

There is a third possibility: host web pages on a mobile device. However, the technology to fully support this option is still in its infancy and is not well explored at present.

6.4.2 Increase Code Abstraction and Reusability

If you have an existing one- or two-layer code architecture, code changes to accommodate the display of pages on mobile devices can be quite difficult, since the presentation, business, and data access logic is intermingled. In a three-layer architecture, the Presentation is abstracted from the Business and Data Access Layers. Thus, when modifying a thin client application to support mobile devices you would only need to make changes to the Presentation Layer.

In addition, using multiple layers of code can provide object reuse when developing fat client applications. For example, a data access object written for one fat client application could be reused by other applications.

6.5 SUMMARY

Mobile client architecture is highly dependent upon the connectivity between the client and server. If the client is always connected, a thin client is often the best solution since the code resides on the server and there is less maintenance and support necessary.

However, if the client is not always connected, a fat client architecture may be necessary. Fat clients, however, are difficult to support and cannot easily be used in applications with huge user populations.

Alternatively, you may use a web page hosting mechanism that allows you to develop web pages for your mobile device and service them even when a back-end web server is not available. Using this option, you can re-leverage any existing web application code you might have and deploy it to mobile devices without having to generate fat client Windows Forms applications.

As coverage and mobile technology improve, we can expect more and more devices to remain connected to networks for extended periods of time. As a result, thin client mobile application architectures may be preferable in the long run.

CHAPTER 7

Client-Server Data Transfer

Evil communications corrupt good manners.

—I Corinthians 15:33

A mobile client may need to download information in order to stay current with back-end system activities. Conversely, a back-end system may need information gathered by a mobile client to be uploaded to it. As a result, it is extremely important to have a reliable mechanism to transfer data between a mobile client and a back-end system.

There are many methods that developers can use to safely transfer data between the two, including HTTP, SOAP, and custom synchronization software (see Figure 7–1).

Each of these methods is best suited for use in different connectivity situations and with different mobile device types. In this chapter, we will discuss each of these methods in more detail.

7.1 HTTP AND HTML

The Web can be thought of as a distributed information system that supports the transfer of hypermedia information between clients and servers. The Web uses HTTP to send and receive information specified using HTML.

A web browser on a mobile client sends a HTTP GET request for information to the web server by specifying a URL. The web server receives this request on port 80, retrieves the specified

Figure 7–1 Data transfer methods

DATA TRANSFER

- **http and html**
- **wap and wml**
- **synchronization software**
- **rda and merge replication**
- **soap and web services**
- **message queues**
- **tcp/ip**

page or file, and responds by sending the requested information in HTML format to the client web browser. The server connection then closes. The client web browser interprets the HTML and displays the formatted page on the mobile device.

The client browser may also submit information to the web server using a HTTP POST request. This allows a mobile client to send information to the web server for processing and data storage.

Web client requests and web server responses consist of multiple series of discrete requests and responses that allow the transfer of data between a mobile web client and web server (see Figure 7–2). This data transfer mechanism is best suited for devices that have an HTML browser and a constant network connection.

7.2 WAP AND WML

WAP is a communications protocol for wireless devices. It works with most existing wireless networks including CDMA, GPRS, and GSM.

WAP is useful to send and receive WML data. WML is similar to HTML and XML but it is used to display web pages on mobile devices such as cellular telephones.

When a client with a WAP browser accesses a web page, the request is initially sent to a WAP gateway. The gateway forwards the request to the HTTP server and receives the response. The gateway then encodes the HTTP response and returns it to the browser as WML (see Figure 7–3).

In WML, developers create "decks of cards." A deck has a URL like a web page and contains one or more cards. Each card represents a small screen of information, such as a menu or a data entry screen. For example, one card may allow the user to type in his/her Zip code to access a weather report; the next card may display the weather report. Generally, each card can only display a few lines of data.

WAP and WML are useful for devices that have a constant connection and a WAP browser, such as cellular telephones.

Figure 7–2 HTTP Request and Response data transfer

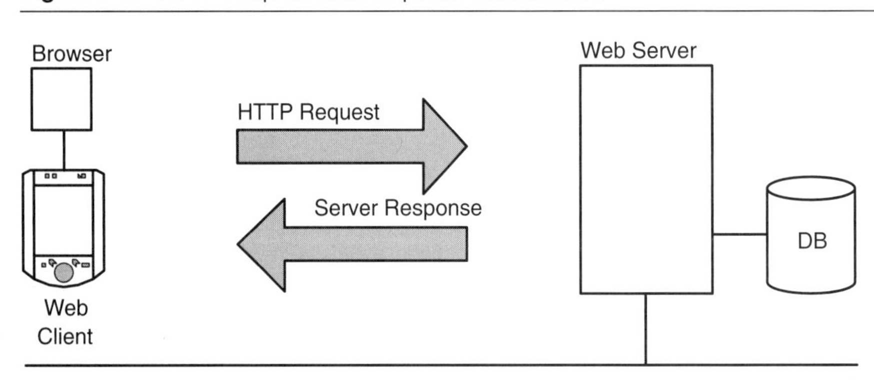

Figure 7–3 WAP and WML data transfer

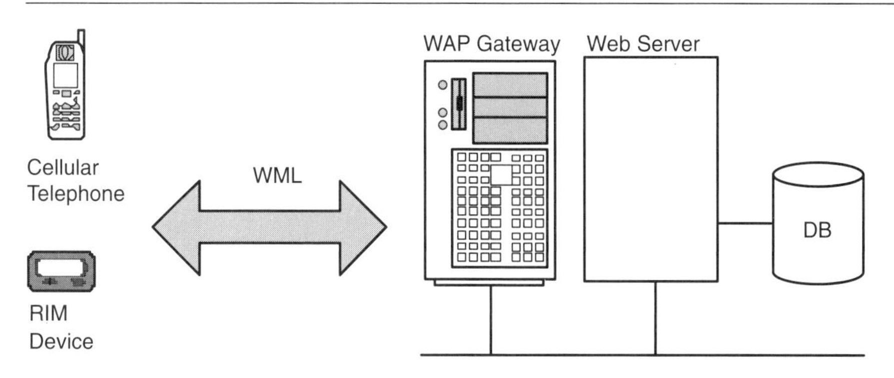

7.3 SYNCHRONIZATION SOFTWARE

There are many off-the-shelf software packages that provide synchronization between a mobile device and a partner computer. One of the most popular for the Pocket PC is Microsoft ActiveSync. This software synchronizes data such as emails, appointments, contacts, tasks, and other information between mobile devices and partner computer applications (see Figure 7–4).

One advantage to Microsoft ActiveSync is that the program generally runs as soon as the user cradles his/her Pocket PC. You do not have to remind the user to upload the data.

You can also use Microsoft ActiveSync to automatically synchronize a Microsoft Access database on a PDA and a Microsoft Access database on the partner computer. However, while this is very useful for small applications, it does not work as well in an enterprise, where multiple users must synchronize with an enterprise class database such as Microsoft SQL Server or Oracle 9i Database.

Microsoft provides a complete Application Programming Interface (API) that allows you to create and register new service providers for use with ActiveSync. A service provider must contain

Figure 7–4 Synchronization software data transfer

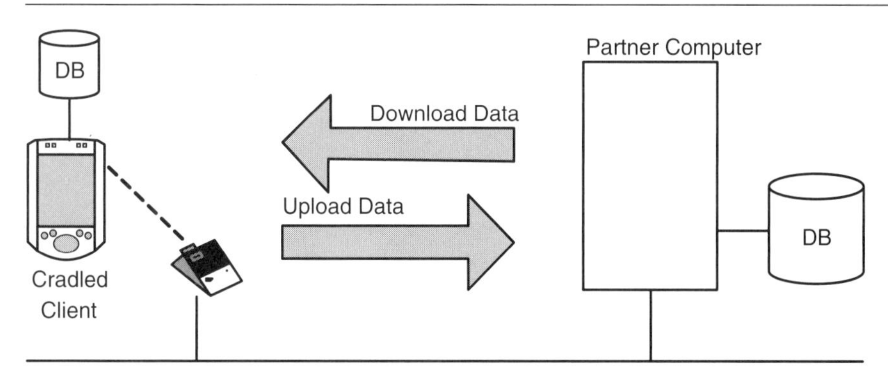

two modules, one that runs on the mobile device and one that runs on the partner computer. These modules work in tandem to determine which data objects should be added, modified, or deleted on the mobile device and partner computer.

7.4 RDA AND MERGE REPLICATION

Many fat client mobile applications provide a store-and-forward mechanism. In the case of a Pocket PC, they may store local data in a Microsoft SQL Server CE database when the user is offline, and then forward the data to a Microsoft SQL Server 2000 database on a server when the user reconnects to the network.

Microsoft provides two mechanisms for synchronizing data between client databases using Microsoft SQL Server CE 2.0 and server databases using Microsoft SQL Server 2000: Remote Data Access (RDA) and Merge Replication (see Figure 7–5).

These two mechanisms have many similarities. Both require installation of the Microsoft SQL Server CE Client Agent on the Pocket PC. In addition, they both utilize the Microsoft SQL CE Server Tools, which run on the web server and are accessed through Microsoft Internet Information Server (IIS). Both mechanisms allow you to download data to the mobile device, edit it locally, and upload changes to the server database. The two mechanisms are described below.

1. **RDA.** RDA allows you to push and pull data between the mobile and server databases. It also allows you to submit SQL statements to be executed on the remote server.

   To pull data from the server, you must specify SQL statements that select data from the various database tables (e.g., Select * from Orders). Microsoft SQL Server CE monitors changes to the data that are made locally so that they can be uploaded later.

   After the user has made changes, you can push the new data up to the server. If there are errors pushing data to the server, the records that were not uploaded are placed into an error table.

   One advantage to using RDA is that no changes need to be made to the back-end Microsoft SQL Server 2000 database. The only action you must take is to install the Microsoft SQL Server CE Server Tools on the web server.

2. **Merge Replication.** Microsoft SQL Server CE replication is based on Microsoft SQL Server 2000 Merge Replication, which uses a publisher/subscriber model. Merge Replication allows you to fully synchronize a Microsoft SQL Server CE database with the server database. You can also specify filters, so that different subsets of data can be published to different local databases.

   Merge Replication requires some changes to the back-end servers. Initially, you need to install the Microsoft SQL Server CE Server Tools. On the database server, you must then create a server publication that specifies which data will be published to other devices. Each synchronized mobile database must subscribe to this publication in order to actually synchronize data.

Figure 7–5 RDA and Merge Replication data transfer

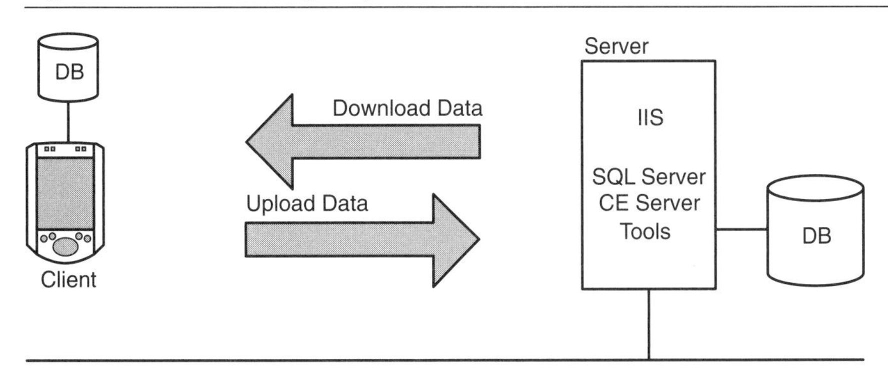

While both RDA and Merge Replication are very useful, there are certain things that need to be considered. For example, you will bypass your server-side business objects by allowing direct database-to-database data transfer. Thus, this should only be used if you include enough business logic in your client application to validate and ensure the integrity of the data you are uploading.

In order to demonstrate the transfer of data between a mobile client and a web server using RDA and Merge Replication, we have created a simple orders application, which allows a user to download a list of products, add new orders locally, and upload the orders to the server. The source code for this application is available for download from this book's companion web site.

7.5 SOAP AND WEB SERVICES

SOAP is a protocol for sending structured data between remote computers over a network. The SOAP specification is defined by the World Wide Web Consortium (W3C). SOAP originally stood for "Simple Object Access Protocol" and was designed as a mechanism for performing remote procedure calls using XML over HTTP. It has now evolved into a more general messaging framework that can run on a variety of protocols. Although it is still called SOAP, as of Version 1.2 the name is no longer an acronym.

SOAP messages containing XML data can be passed from a client to the server. Because SOAP is an open standards-based protocol, it can also be used from any operating system and can provide interoperability among devices running different operating systems.

The SOAP messaging framework defines rules for constructing and processing SOAP messages. Each SOAP message is an XML document consisting of a SOAP Envelope. The Envelope must contain a Body element, which contains the message payload or error information. If an error occurs, the Body should contain a Fault element with any error messages that were generated by the method call.

The Envelope may also contain a Header element used to pass additional information that is not part of the main method call. This may include the message priority, expiration time, or authentication information. If it is present, the Header comes before the Body element.

SOAP has become a useful tool for companies exchanging data over the Internet as a result of its ability to transmit XML data. It is a more modern and less costly alternative to the traditional Electronic Data Interchange (EDI) protocol that has been used by many companies over the years.

SOAP is also beneficial in mobile application development. Although it requires a network connection between the client and the server, it can be used in a store-and-forward scenario if you store data on the client device and forward it using SOAP once the network connection is established (see Figure 7–6).

One disadvantage to using SOAP is that messages can be quite large, since all information is encoded using XML. However, the turnaround time of a request can be quite high, so nevertheless it is probably better to make one request that returns a large amount of data than to make multiple requests that return a small amount of data.

You may often hear the term "web service" used almost interchangeably with SOAP, but the two terms are not exactly synonymous. A web service is a software component that can be called from any platform or operating system over the Internet. Web services are described using XML and WSDL. They communicate using SOAP and can be located by other systems using the Universal Description, Discovery, and Integration (UDDI) protocol. These are discussed in more detail below.

WSDL. The Web Services Description Language (WSDL) provides a specification for describing a web service in XML. The description contains information about the messages, data types, and protocols that the web service supports.

A WSDL document is basically an XML schema document describing a web service. It contains the following items:

- Service Name—the name of the web service
- Port—the actual location, or URL, of the web service
- Binding—the transport and encoding protocols used by the port
- Port Type—the operations supported by the web service
- Messages—the request and response messages for each operation

UDDI & DISCO. UDDI is a standard used in order to discover web services. The UDDI specification defines a SOAP-based mechanism for locating web services defined by WSDL.

Developers can register their web services in a central directory using UDDI. Two types of information are registered with UDDI. One is the abstract technical model of the web service, known as the *tModel.* The tModel describes the business service offered by the web service. The second type of information is the actual service implementation, and is known as the *business entity.*

Figure 7–6 SOAP data transfer

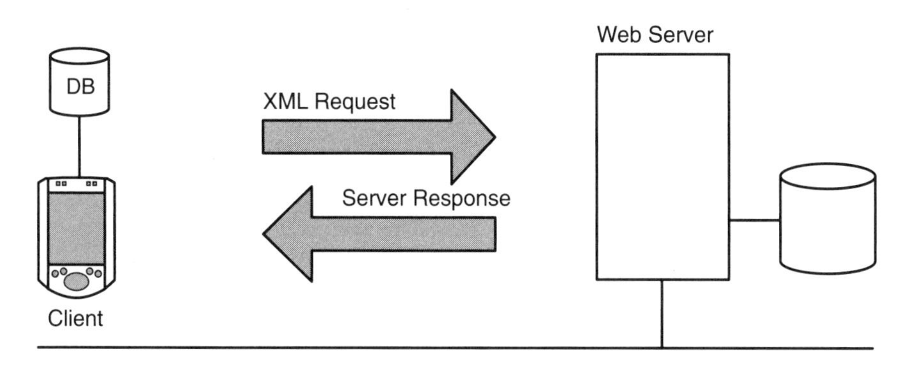

The tModel contains the company name, a user-friendly name, and a description of the service offered. It contains a globally unique identifier (GUID) for the service and the URL of the service's WSDL file. It may also contain information about the web service's category.

In addition, web services can be published and discovered using a Microsoft proprietary mechanism called DISCO. A web service may publish a discovery file, which is an XML file with the .disco extension. This discovery document can then be deployed to a web server to enable discovery of the web service.

One advantage to using web services is that you can releverage the business objects you already have running on your server. Therefore, if you have already invested time and effort in creating business objects, you can expose those objects through the use of a web service. This allows mobile devices to call the objects on your server whenever they are connected to the network.

In order to demonstrate the transfer of data between a mobile client and a web server using an XML web service, we have developed a simple web service that performs arithmetic calculations called the SOAPCalculator. Information gathered by a user on a mobile device (e.g., two numbers) is sent to a server for calculation through the web service and the result is passed back to the mobile client. For example, "2+2" is entered on the mobile client. The addition operation is calculated on the server and the result, 4, is returned to the mobile client. The source code for this application is available for download from this book's companion web site.

7.6 MESSAGE QUEUES

Message queues can be used by mobile applications that are partially connected and use a store-and-forward mechanism. For example, an application may allow the user to upload data from a mobile device to a server. If there is no connection initially, each user update can be put into a message queue. Once a network connection is established, the client message queue software can forward each message to the server message queue and await a response.

Figure 7–7 Message queue data transfer

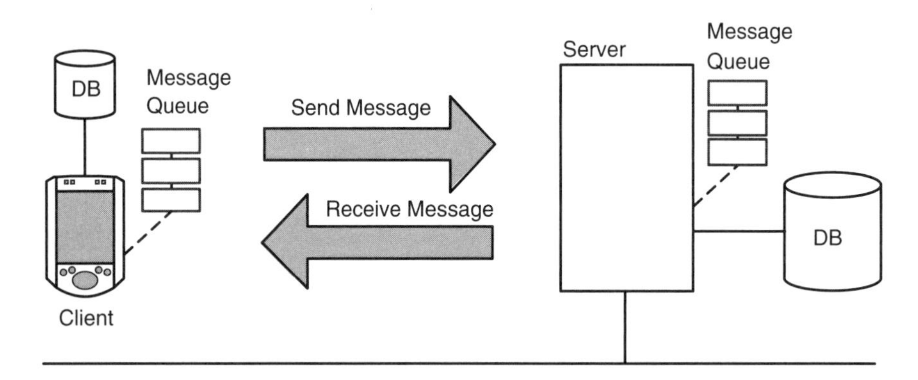

A mobile application can also have an event handler that is triggered when a message response is received from the server. Depending on what the response is, the mobile application can take appropriate action including displaying a message to the user, updating the database or performing other clean-up tasks. Microsoft provides a version of Microsoft Message Queue (MSMQ) for the Pocket PC.

One advantage to message queues is that they allow you to place a message in the queue and essentially not worry about it. The queue software will repeatedly attempt to send the message to the server until it is successful (see Figure 7–7).

However, one disadvantage to using message queues is that you must manage the queue on the server. The use of message queues also generally requires more setup and administration than most of the other data transfer mechanisms we discuss in this book.

7.7 TCP/IP

TCP/IP provides a sequenced, reliable, two-way communication mechanism when a mobile device is connected to the network. The TCP/IP server application listens on a specific network port for incoming client requests. The client application initiates communication with the server by sending a request packet. When the server receives the request, it processes it and responds. After this initial sequenced message exchange, the client and server can exchange data (see Figure 7–8).

To exchange data, the server application creates a new socket, which it binds to a particular IP address and port. Then the server waits for incoming requests. When a new request is received, it spawns a new worker thread to process that request and send a response.

The client application also opens a socket and connects to the server on the given IP address and port. The client then sends and receives messages over the network as arrays of bytes.

In order to demonstrate the transfer of data between a mobile client and a server using TCP/IP, we have developed a simple communication server that receives data from a remote

Figure 7–8 TCP/IP data transfer

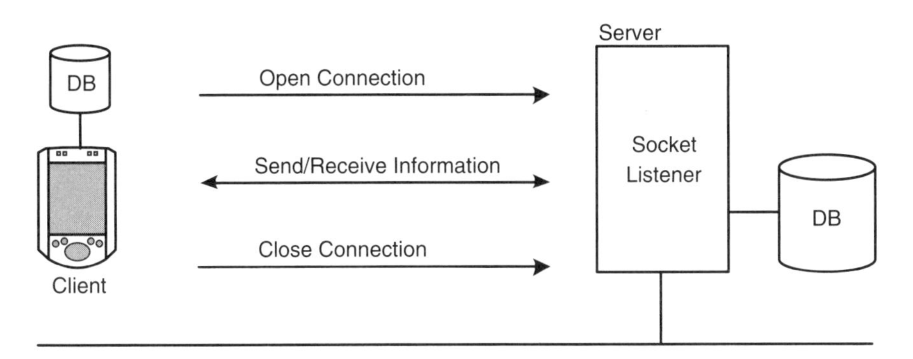

client and echoes it back. This communication server listens for client connections and starts up new threads as needed to handle the client communications. Each thread is an asynchronous TCP/IP listener. When a client program connects to the communications server, the server creates a new thread and a new state object to hold the data sent by the client. Once the client enters an end marker string, the data is echoed back to the client. The source code for this application is available for download from this book's companion web site.

7.8 SUMMARY

This chapter described several methods that developers can use to transfer data between a mobile client and a server. Some of these methods are suited for applications with a constant connection, while others are useful for store-and-forward applications, and some can be used for both.

HTTP and WAP can be used to service web page requests from mobile devices that have web or WAP browsers that can render HTML or WML.

Microsoft ActiveSync can be used to handle standard synchronization of emails, appointments, tasks, and contacts and is probably the most common type of synchronization used by Pocket PC users.

RDA and Merge Replication can be used to synchronize data between Microsoft SQL Server CE and Microsoft SQL Server 2000 databases. This is useful in many applications but can also cause problems, since it bypasses server-side business objects.

If you want to leverage your existing server code, SOAP can be used to expose server-side business objects as web services so that they can be called by mobile applications across the network.

Message queues can also be used to send data updates between client and server applications. However, this does require queue setup and administration.

Finally, TCP/IP can also be used for two-way communication between mobile clients and servers.

CHAPTER 8

Mobilizing Existing Application Architectures

What's gone and past help, should be past grief.

—Shakespeare

The development of mobile applications not only includes creating a mobile client that communicates with your back-end systems, but often requires modifying and extending your existing application architecture.

In this chapter, we initially look at how the architectures of modern enterprise web applications have evolved. We then explore some of the modifications and extensions that will have to be made to these architectures in order for them to accommodate new mobile applications.

8.1 EVOLUTION OF ENTERPRISE ARCHITECTURES

In the following sections, we look at the evolution of enterprise architectures as well as some of the different architectures that may be found within any typical enterprise.

8.1.1 Client-Server Architecture

Traditional enterprise architectures are composed of multiple large servers that run a combination of applications critical to the enterprise. In a bank, this may be a credit and debit system. In a brokerage, this may be a trading system. In a manufacturing company, this may be an inventory system. There are also billing and payment systems as well as order and fulfillment systems. Other such systems include customer relationship management systems, which manage customer information, and human resources systems, which manage employee information.

These application systems are variously called "back-end," "back-office," "legacy," or "mission-critical" systems since they are critical to an enterprise and serve as the business computing power behind an organization.

Many of these back-end systems are mainframes or servers that run older operating systems, languages, and technologies. The management and operation of these back-end systems were traditionally carried out by a variety of users, such as administrators, help desk personnel, operators, and other back-office personnel. These users typically used "dumb" terminals that were connected to the mainframes or servers.

With the advent of the Web, however, many of these back-end server systems were re-engineered to allow customers and front-end users to access information and services through web sites. In doing so, a form of customer self-service has become the norm (see Figure 8–1). This has been beneficial in several ways, to both service providers and customers or end-users.

For example, a few years ago, it was sometimes difficult to obtain your checking or savings account balances quickly, because the information was stored in back-end systems. Banks typically provided monthly statements, while day-to-day queries about balances, payments, etc. were handled through ATMs, tellers, or telephone operators. Today, many banks have placed front- and middle-tier applications in front of their back-end systems, allowing users to view checking or savings accounts on the Web at almost any time.

Through the Web's influence, valuable legacy applications have been given a new lease on life since they have been modified to allow web access while retaining their core functionality.

8.1.2 Web-Enabled Service-Centric Architecture

Traditional back-end systems and applications provided customers and users with a set of application services that could be viewed over the Web. These application services, however, were not typically implemented or deployed all at once, but grew over time. As a result, users were often identified on a service-by-service basis and there was very little sharing of user identity or personal information between multiple services. This type of architecture may be called a *service-centric architecture* (see Figure 8–2).

Figure 8–1 Web and client-server architectures

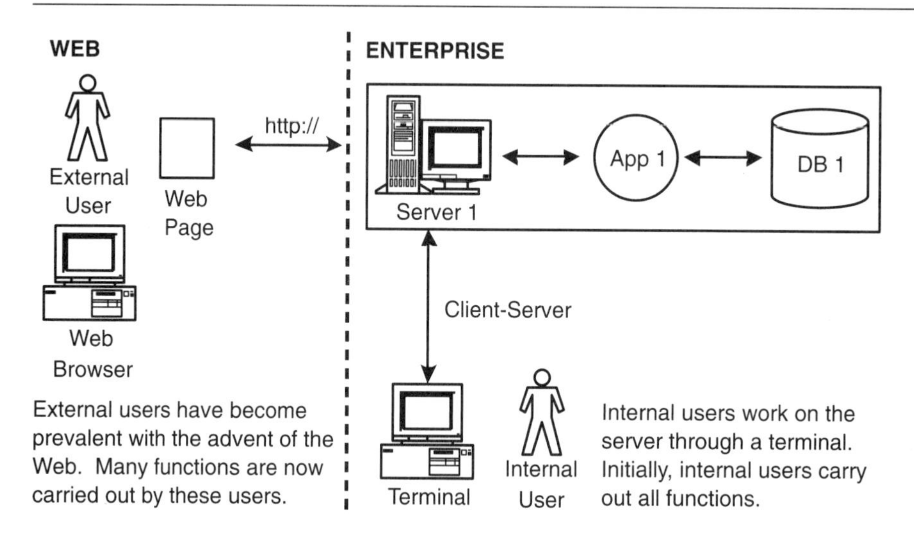

Figure 8–2 Service-centric architecture

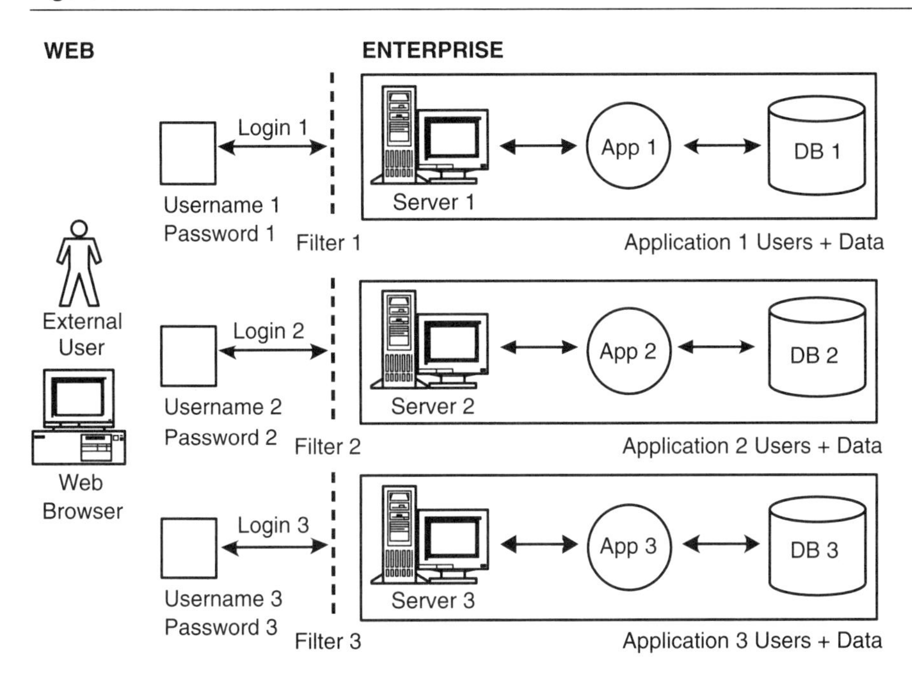

In a service-centric architecture, a service provider offers multiple services. Each server and application has its own user base and information and is responsible for its own security. There is little sharing of resources between applications and external users must log in to each service separately.

For example, a bank may offer users access to several different accounts (checking, savings, etc.) but require that a user have several usernames and passwords and log in to each account separately. Another example is a company that offers users access to order and billing services on two separate systems, requiring them to have two different usernames and passwords.

Thus, companies that offered multiple services often forced users to log in to each service separately. This was cumbersome for users, who needed to remember or maintain lists of usernames, passwords, and account IDs. In addition, this architecture did not lead to a highly integrated or personalized browsing experience since there was little or no connection between the individual application services.

8.1.3 User-Centric, Single Sign-On Web Architecture

The ability to implement user-centric, single sign-on web architectures directly influences the business value that a user equates to his/her relationship with a service provider. This can be attributed to many factors such as service level, service consistency, and how well a user perceives that there is a differential in doing business with the service provider.

In the case of a bank, for example, a user should be able to access all his/her accounts online using one username and password. In addition, a user does not and should not care about individual application service details.

A user-centric architecture achieves this by providing a unified view of the multiple services provided by the company. In this architecture, each server and application shares some common resources (such as the user database and security), while each application handles its own specific functionality. Security is typically placed in a system called the *authentication authority* that is trusted by all services. As a result, users can log in once and be granted an authentication token that is recognized by each application and permits them access to multiple services (see Figure 8–3).

Figure 8–3 User-centric architecture

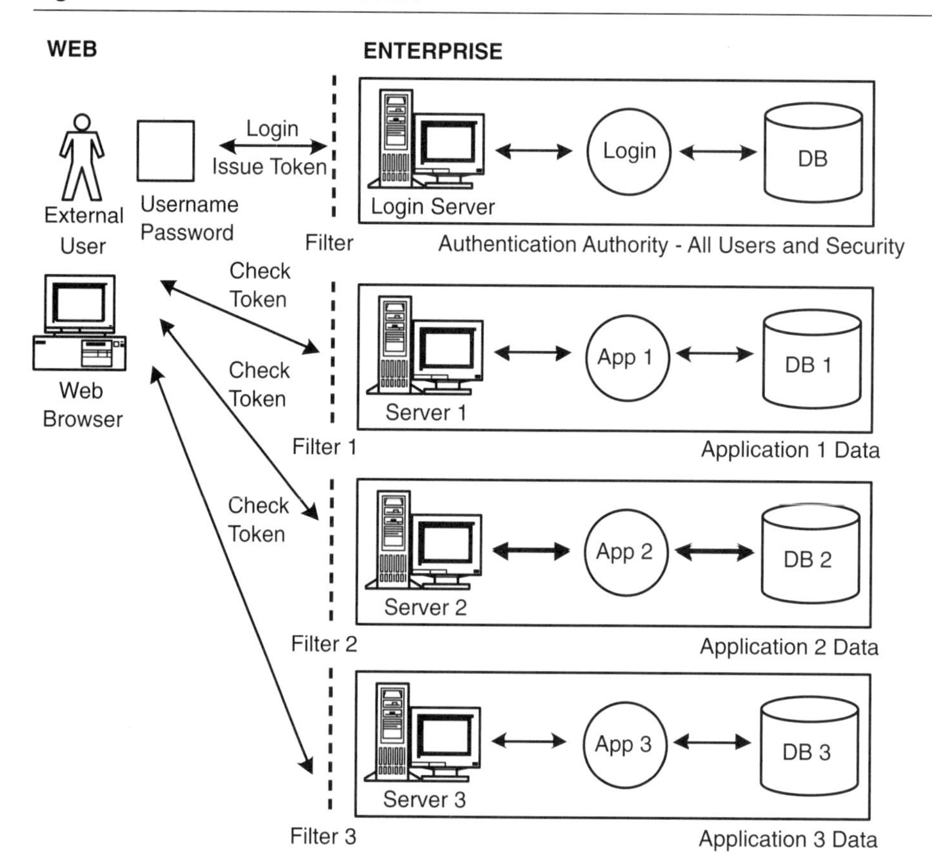

8.2 ANATOMY OF AN ENTERPRISE WEB ARCHITECTURE

Recently, the widespread use of mobile devices has led to the need to integrate mobile applications into existing enterprise web architectures. In order to provide a more concrete discussion of how we might do this, we will initially characterize and present the anatomy of a typical modern enterprise web application.

8.2.1 Architecture

A large-scale, enterprise web architecture is typically composed of a set of servers that are placed in three tiers: the Presentation Tier, the Application Tier, and the Database Tier (see Figure 8–4).

The Presentation Tier consists of a set of web servers that host public, private, and administration web pages, such as login, enrollment, and application-specific pages. These servers may also host web services.

For example, the Presentation Tier web server can display a login page to a user. Upon user data entry and form submission, the login data is passed to the authentication objects in the Application Tier for verification. The Presentation Tier web servers are protected from unauthenticated user access by an authentication filter. They are hosted in the DMZ and are sandwiched between an Internet firewall and an intranet firewall.

The Application Tier application servers host common authentication, user, access control, data access, and application-specific objects. For example, an access control database containing user security information is accessed through objects residing on these application servers. The application servers may also be protected by an authentication filter and any other filters necessary to prevent unauthorized access. They are typically hosted within an enterprise and are shielded by a firewall.

The Database Tier servers host the access control database, which holds the user profiles, roles, and entitlements tables, among others. The database servers are also typically located within an enterprise behind a firewall. They are tightly restricted as they hold information critical to both the enterprise and the user.

Note that in Figure 8–4 we only show a single server per tier. In practice, there can be numerous presentation, application, and database servers. This architecture is both horizontally and vertically scalable. In other words, you can add more servers and also more powerful servers.

8.2.2 Logical View

Multiple products and technologies from a variety of vendors exist to assist in the development of enterprise web architectures. Figure 8–5 shows a logical view of the major architectural components. Below, we describe some of these components in more detail.

Microsoft technologies. Enterprise web applications can be implemented on Microsoft Windows 2003 servers running Microsoft IIS and Microsoft SQL Server 2000 as the platform for access control, authentication, enrollment, and personalization. Web pages are built

Figure 8–4 General enterprise web architecture

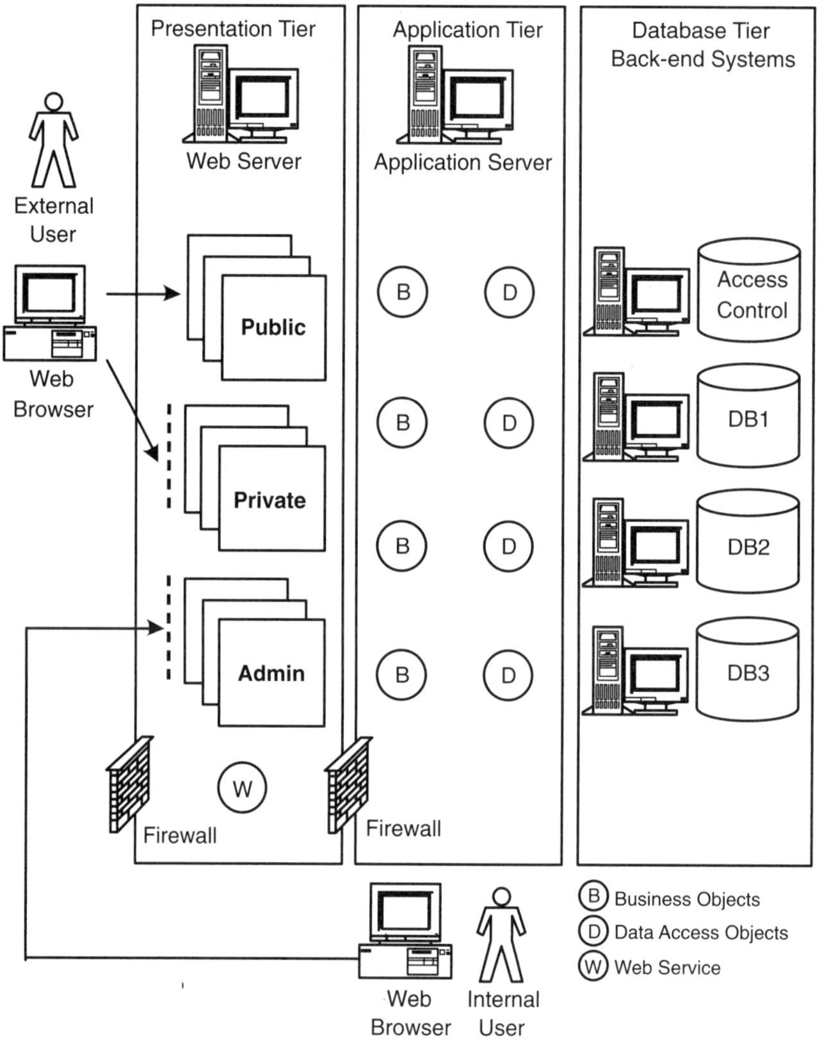

using ASP.NET Web Forms. Business logic is implemented through VB.NET and C#.NET application objects that handle the reading and writing of data from the Application Tier to the back-end systems. Microsoft SQL Server physically houses the access control database.

Oracle & Sun technologies. Enterprise web applications can also be implemented using a Microsoft Windows or a UNIX server running Oracle 9iAS (OC4J) and Oracle 9i DB as the platform for access control, authentication, enrollment, and personalization. Web pages are presented in JSP and HTML. Business logic is implemented through

Figure 8–5 Logical view of major architectural components

Presentation Web Pages ASP/HTML/XML JSP/HTML/XML	Login Page	Enrollment Page	Application Page	Application Page
Business Application Objects VB/VC++/COM/.NET JAVA/J2EE/EJB	Authentication Object	Enrollment Object	Application Object	Application Object
Platform	Microsoft Technologies (e.g., Windows, IIS, .NET Servers) Other Technologies (e.g., UNIX, Apache, Oracle OC4J)			
Network	TCP/IP, SNA			
Back-end Systems Databases	Users Roles Entitlements	DB 1	DB 2	DB 3

Java/J2EE/EJB objects that handle the reading and writing of data from the Application Tier to the back-end systems. An Oracle database physically houses the access control database.

Other technologies. There are many other products and technologies that can be used, including IBM WebSphere, Netegrity Site Minder, Citrix NFuse Elite, and IONA's portal. For further product details, please consult the specific vendor's product documentation.

Out-of-the-box product sets are somewhat of a two-edged sword since they have relative strengths and weaknesses that can be beneficial or detrimental to your particular deployment. An out-of-the-box product is often quicker to deploy and often has the benefit of being supported by a large vendor or corporation. However, these products can be very tightly coupled, which does not lend well to heterogeneous or complex application development environments. Furthermore, the look and feel is often "canned," which may not be acceptable.

As a result, you may well encounter a situation where an out-of-the-box product that is used to generate an enterprise web application does *not* easily support mobile devices or integration with back-end systems. Our opinion is that the "best" solution is often to use a mix of loosely coupled, out-of-the-box products along with customized code.

8.3 CONSIDERATIONS WHEN MOBILIZING EXISTING APPLICATIONS

Most enterprise web architectures were implemented long before mobile applications existed or were considered. As a result, the need to accommodate new mobile applications with existing applications can have far-reaching effects on the existing applications and servers.

In the following sections, we will discuss several important considerations that need to be addressed when mobilizing existing applications. There is no implicit order in the considerations. However, typically, they will all need to be considered at one stage or another.

8.3.1 Architecture

When we mobilize existing application architectures, we are attempting to add mobility components to the overall architecture (such as application code, mobile clients, and servers). In doing so, we are extending the existing enterprise application services to encompass mobile users.

The ability to extend three-tiered web architectures to accommodate mobile applications depends on three major considerations. These considerations are illustrated in Figure 8–6 and described below.

1. **Connectivity.** A major consideration when mobilizing your existing enterprise applications is the type of connectivity that you may utilize. In other words, will the mobile user be always, partially, or never connected?

   If the mobile user will only be partially connected, you will probably need to develop a fat mobile client that primarily works in standalone mode, and only connects to the server for synchronization. In this case, the server's existing web pages may only be minimally affected since the mobile client will require a new application. Much of the connectivity effort will involve the development of a web service that allows communication between the mobile client and the server's business object.

   If the mobile user will always be connected, however, the server's Presentation Tier will require a new set of pages that can be displayed properly on a thin mobile client through a web or WAP browser.

   The connectivity type also has an effect on the existing enterprise's authentication, administration, and security. Each of these factors will be considered in more detail later.

2. **Level of abstraction.** If the current level of abstraction in your existing enterprise is quite good (i.e., there is a clear separation between presentation, business, and data access logic), adding mobile applications potentially will be easier because you may be able to reuse the existing business and data access application components.

   For example, one database may hold customer personal data while another database may hold customer financial account data. In order for your new mobile application to extract, display, and update the information in these systems, you might have to write quite a lot of code. However, if the business and data access logic already exists, it may be possible to reuse the server code with minor or no modifications.

   If your existing enterprise applications are not well-abstracted (i.e., the presentation, business, and data access logic is intermingled), potentially a great deal of work may have to be done. We recommend that you consider this before you attempt to mobilize any existing enterprise applications.

Figure 8–6 Enterprise web architecture with mobile extensions

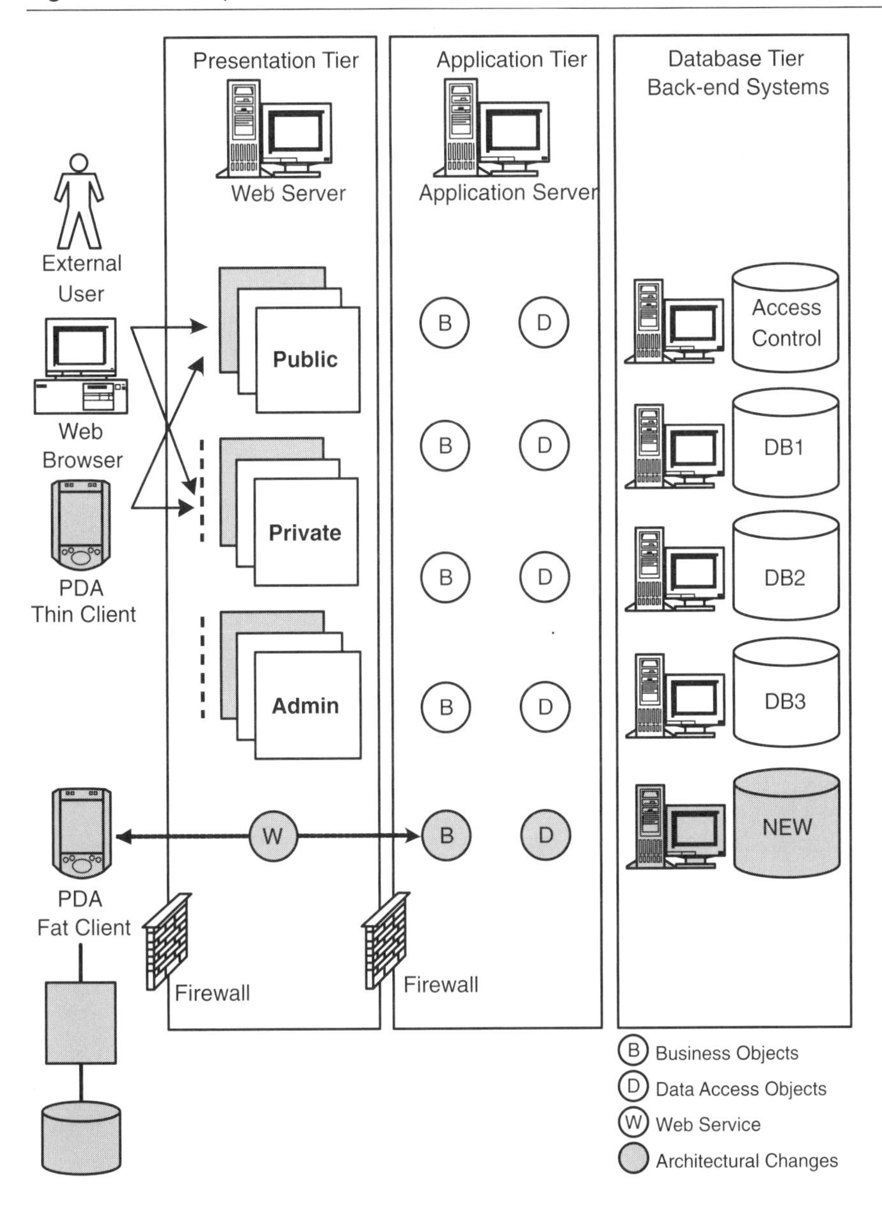

3. **Scaled-down version of existing application or new application.** A third consideration in mobilizing your existing enterprise application is what you are actually mobilizing. If you are mobilizing an existing enterprise application and your connectivity is partial, the fat client you develop will probably be a scaled-down version of the full server

application. For example, you may simply provide the data-gathering aspect of a server application on the mobile device and leave the data analysis aspect on the server. Using this method, the new mobile application simply acts as a new channel for data entry into the server application.

However, if you intend to build a new mobile application with business logic that is not currently supported by the server, you will also need to build the server's new presentation, business, and data access logic along with the back-end system interface.

8.3.2 Users, Roles, and Entitlements

The ability to manage users in different capacities is fundamental to the success of enterprise web applications. In order to do so, users must be defined, assigned a role, and granted or denied access to the application through application entitlements. This is carried out initially during enrollment. Subsequently, when a user logs in and is authenticated, he/she is recognized and placed into a specific role-based group and granted or denied access to specific applications (see Figure 8–7).

Users. Broadly, there are two types of users:

1. **Unregistered users.** All users are typically allowed to browse the public pages of the web application without explicit login or enrollment. Unregistered users remain anonymous and the web application usually provides no information on or tracking of these casual visitors.
2. **Registered users.** Registered or enrolled users are known to the web application and can access secured portions of the web application by logging in. A registered user has a profile stored in a database, which allows the application to personalize the user's viewing experience. In other words, depending on who you are, a different view of the application can be rendered. Each registered user is typically able to modify his/her own user profile and customize sets of web pages based on information contained in his/her user profile record. Certain privileged users may also carry out additional functions on other user accounts.

Role-based groups. Registered users are typically placed in a role-based group. Groups are a convenient way of categorizing users based on their role.

1. **External service-enrolled users.** External service-enrolled users (e.g., customers) have registered and have been placed in a specific group that allows them to use the application. For example, an enrolled user may modify his/her profile information and conduct application functions such as banking or trading.
2. **Help desk personnel.** Help desk personnel provide assistance to all users of the enterprise web application. Help desk personnel may need the ability to

Figure 8–7 Users, role-based groups, and entitlements

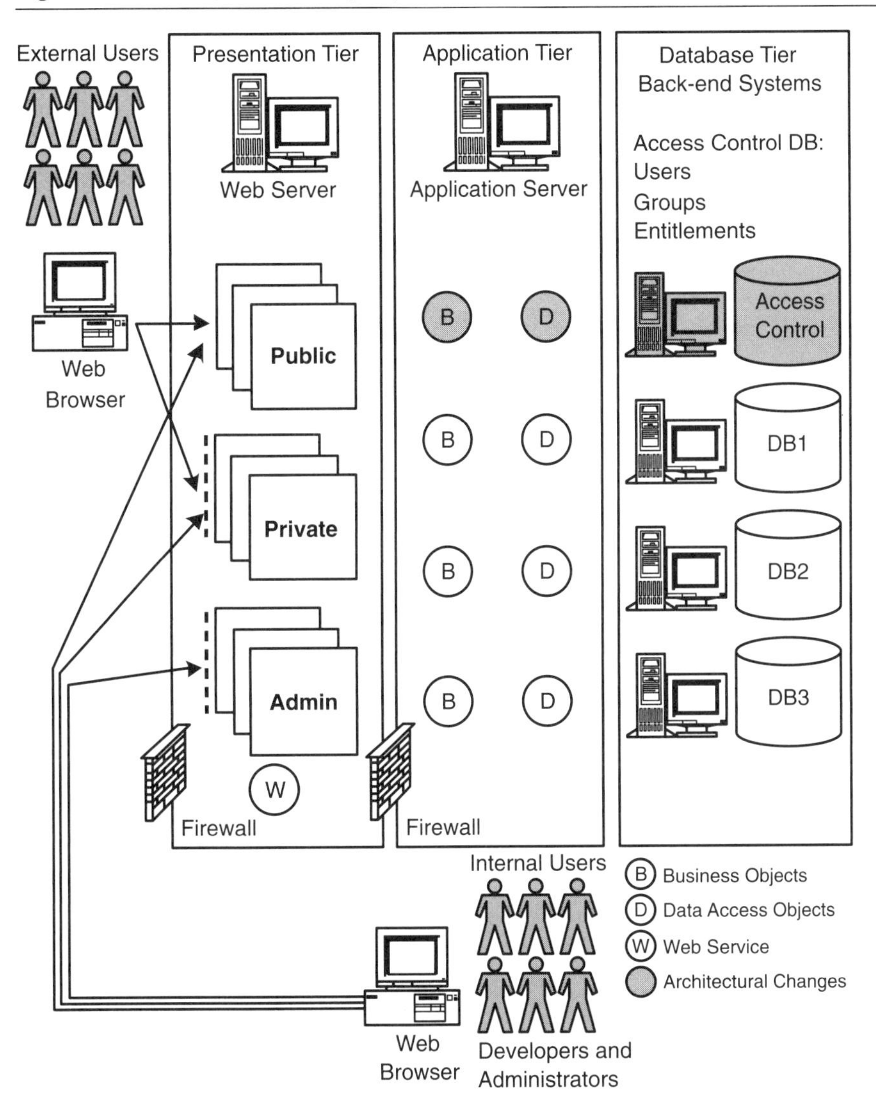

impersonate users (such as external service-enrolled users) and be able to carry out functions on their behalf.

3. **System administrators.** The highest-level user in an enterprise web application is typically the system administrator. Each system administrator will have the ability to add, modify, and/or delete help desk personnel and service-enrolled users.

4. **Others.** Many other groups can also be defined. For example, in a news and media enterprise, content authors may be placed in a group that is allowed to release documents and other web page content to the web application. In a bank, employees such as financial advisors might form a group. Developers may form another group and computer operators yet another.

Entitlements. Once a user is authenticated, he/she is granted access to various applications. Entitlements define what a user can see and do in these applications. Typically, there are two types of entitlements: functional entitlements (what a user can do) and data entitlements (what a user can see). Entitlements are stored as fields in an access control database and permit personalization and customization of the web application.

When you attempt to mobilize your existing enterprise web applications, you will need to consider users, groups, and entitlements carefully. The number of users may not differ greatly from the existing user population, nor may the group membership. However, what may change significantly are the entitlements a mobile device user is granted.

You may decide, for example, that one particular mobile device is not secure enough or powerful enough to handle the existing application's functionality. You may then decide to prevent a user from seeing or using this functionality on this mobile device. In order to do so, a detection of the device accessing the application will need to be implemented. Once a device type is detected, the application entitlements are consulted to see what the mobile user is allowed to do and see. Access to the server application web page and object functionality may be disabled as a result.

8.3.3 Presentation Tier

The Presentation Tier in an enterprise web application consists of web pages that present and gather information from a user. Web application pages can consist of HTML, XML, WML, JSP, ASP, or ASP.NET pages. Advanced-level enterprise web applications may also render these pages dynamically or generate them via code running behind the scenes (see Figure 8–8). The following functionality is typically found in the Presentation Tier:

- Web pages
- Application page rendering functionality
- Logic to verify form entry (e.g., JavaScript)
- Read and write access to objects in the Application Tier
- Web Services

Reworking of the server-side Presentation Tier depends on what the connectivity of your client will be, since this in turn determines the type of client the server has to support. The two types of clients and their related support issues are described on the next page.

1. **Thin client.** With a thin client, there are typically a number of Presentation Tier changes needed, since the mobile device will not have an application running on it and the web page logic will reside on the server. In this case, you may need to:

   - **Support as many users as possible.** The application should be designed to handle the most users possible. For example, a web application with large graphics may look beautiful but if the mobile users only possess low-speed modem lines, the performance may not be satisfactory. Thus, the full range of users must be kept in mind, including those with high-performance and much lower performance systems.
   - **Support as many mobile device types as possible.** This may result in a lowest common denominator approach. If you have to support many mobile device types—some with limited communication speed—the web pages may have to be very simple.
   - **Support as many browser/display types as possible.** Again, this may result in a lowest common denominator approach. If you have to support many browser/display types, the web pages may have to be relatively simple.
   - **Support browser neutrality and minimum user system requirements.** The application should support browsers from several major vendors (e.g., Microsoft Internet Explorer and Netscape Navigator). If browser-specific components are utilized, equivalent functionality should be provided for any other supported browser. For example, an ActiveX component specific to Microsoft Internet Explorer should not be deployed without providing equivalent support for Netscape Navigator.

   The basic problem to be solved here is how to display a set of large, existing web pages (with images and pop-up windows) on small mobile devices so that the applications are tolerable to use. In this case, it is likely that you will need to develop several sets of web pages on the server, as follows:

   - Original web site pages (public, private, administration)
   - Mobile version of the original web site pages (public, private, administration)

   The design of the original web pages and mobile client web pages should follow the user interface and mobile client application development guidelines discussed in detail in Chapter 5 and Chapter 6.

2. **Fat client.** With a fat client, there are typically very few Presentation Tier web page changes needed, since the mobile device will have its own new application pages and functionality.

   However, you may need to implement a web service on the server that can forward requests to your existing business objects. This is placed in the Presentation Tier of a three-tier architecture because it is hosted by the web server (e.g., Microsoft IIS) and may need to be accessible to users over the Internet.

Figure 8–8 Presentation Tier in an enterprise web architecture

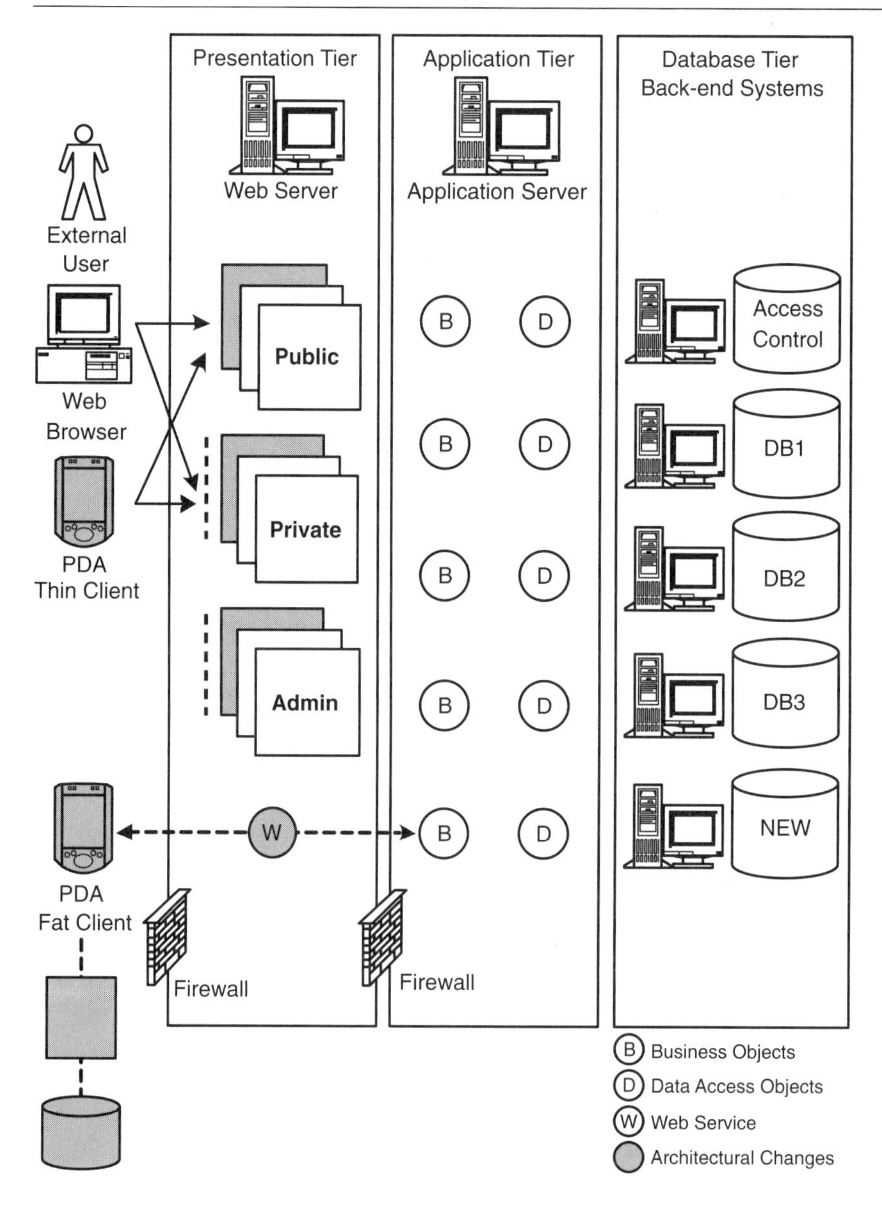

8.3.4 Application Tier

A set of business and data access objects and components is typically implemented on application servers in the Application Tier. They provide the following functionality:

- Authentication
- Enrollment to various services
- Retrieval and storage of user and back-end system data
- Application-specific business functionality
- Session management (optional)
- State management (optional)
- Caching (optional)

The business and data access objects and application services provide a level of data abstraction and aggregation between the front-end and back-end tiers of an enterprise architecture. They also allow front-end client applications to quickly and easily retrieve back-end data without a detailed knowledge of the underlying data store structure (see Figure 8–9).

If there is good abstraction in the existing enterprise web architecture, few changes will typically be made in the Application Tier when mobile fat clients are introduced except perhaps to verify and test the introduction of a set of new web services in the Presentation Tier that communicate with the existing business objects.

However, if there is insufficient abstraction in the existing architecture, a great deal of work may be required to introduce an Application Tier and an accompanying set of business application services. For example, if the existing web page and business logic is intertwined, it may not be easy or possible to reuse the business code when a mobile application that uses smaller pages is introduced.

8.3.5 Database Tier

In all but the smallest of architectures, there is a need to be able to uniquely identify the user, place the user in a role-based group, and base the user's entitlements on the user's relationship with the individual service provider. The security data needed to do this is typically stored in an access control database (see Figure 8–10).

An enterprise web application's access control database is composed of several tables, each of which is described later in this section. For convenience, we have assumed that a single database holds all the information. However, each table can be in its own database and each database can even be on a different platform. For example, you could place the User Profile table in a Microsoft SQL Server database and another table in an Oracle database, etc.

Figure 8–9 Business Tier in an enterprise web architecture

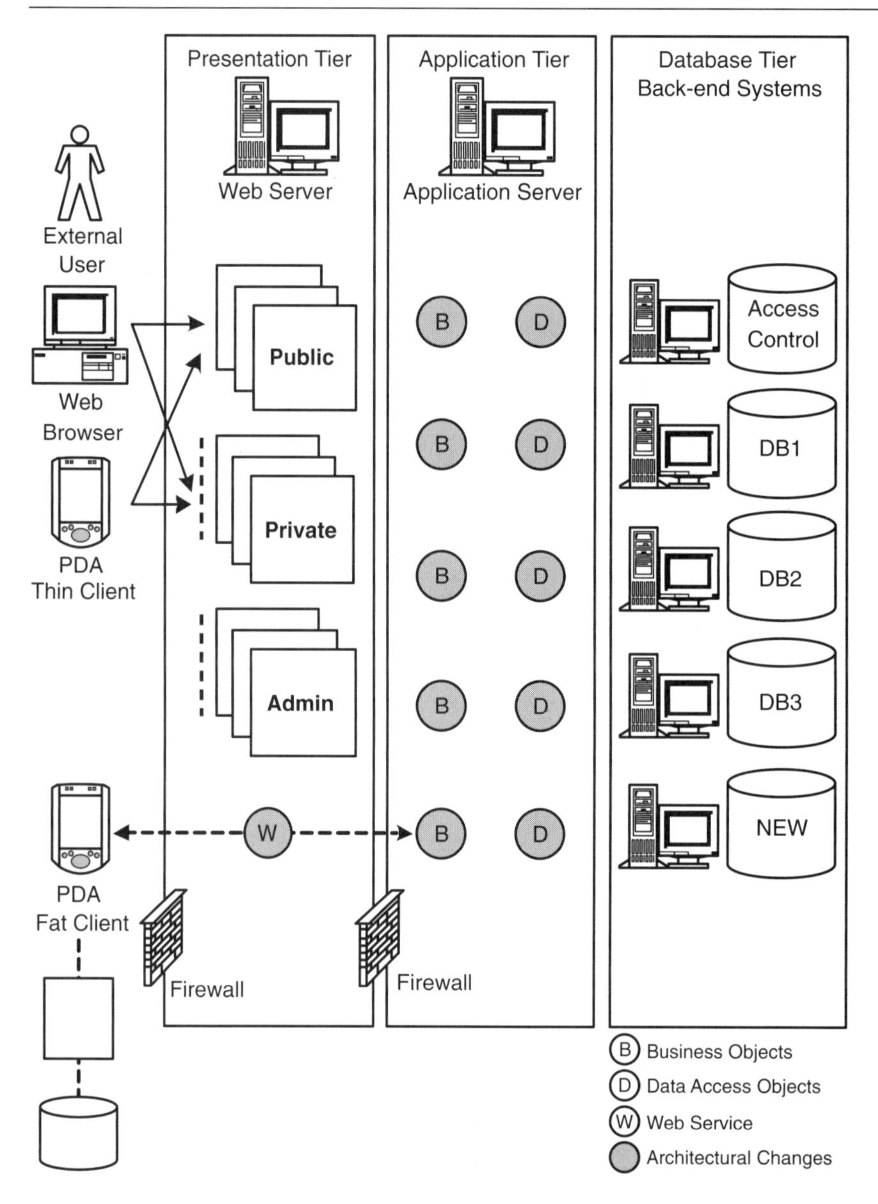

Figure 8–10 Database Tier in an enterprise web architecture

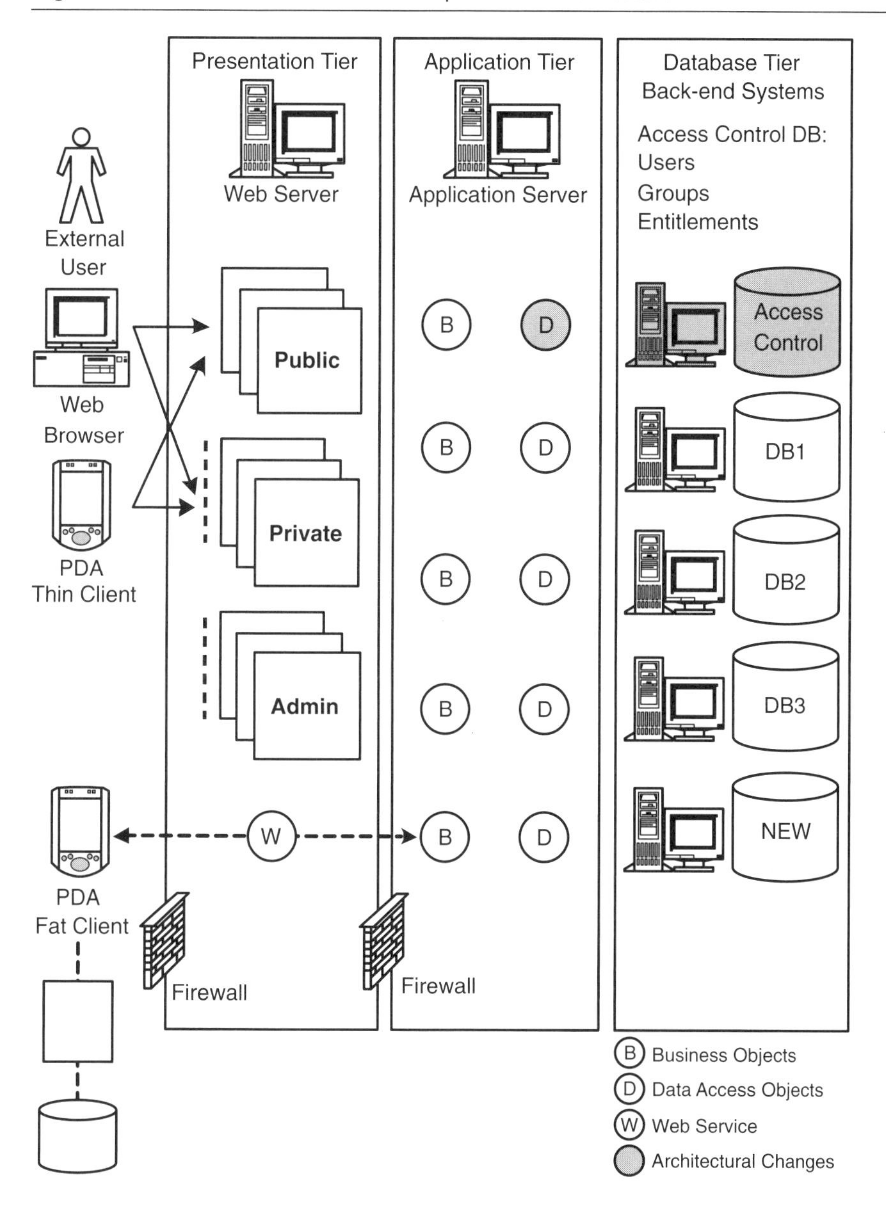

Access control table. One of the key principles behind large-scale web applications is the ability to have the user identity held in a separate table and connect this independent identity to any number of other applications. Some of the fields include:

- UserId (primary key)
- Username
- Password
- Disabled
- RoleId (foreign key)

Role-based groups table. Users may be organized into a variety of role-based groups. These groups are created and maintained by an administrator. Some of the fields include:

- RoleId (primary key)
- RoleName
- RoleDescription

Application entitlements table. The application entitlements table describes the entitlements each user has for each application. Some of the fields include:

- AppId (foreign key)
- UserId (foreign key)

Applications table. The applications table describes all applications. Some of the fields include:

- AppId (primary key)
- AppName
- AppURL
- AppDescription

Personalization table. Personalization includes the ability to uniquely identify a user according to information stored in the database and perform custom/personal tasks based on that information. This might include personalized home pages, preferred styles, greetings, languages, news, help, account, and advertising information. Some of the fields include:

- UserId
- HomePageURL
- PreferredStyle

User profile table. The user profile table holds the core information for each user. In large-scale web applications, the total number of users can be expected to be in the millions. Each user has a set of attributes that identify him/her and allow personalization to occur. The number of attributes and the contents of each attribute can be expected to play a large role in determining the size of the table. Some of the fields include:

- ProfileId (primary key)
- UserId (foreign key)
- Firstname
- Middlename
- Lastname
- Address1
- Address2
- City
- State
- Zip
- Country
- Telephone
- Email

If the existing web architecture has an Application Tier that already interfaces to the Database Tier, changes to the access control database will be probably be minimal. Except within the entitlements table, which you may impose certain restrictions on mobile users for security reasons.

8.3.6 Existing Back-end Systems

In a three-tiered architecture, the addition of a set of mobile users and mobile devices will probably have minimal impact on the existing back-end systems and applications since the business objects and data access interfaces to these systems will already be implemented in the Application Tier. However, back-end application testing time may be required to ensure end-to-end operability. In addition, back-end application re-certification may need to be considered and performed (see Figure 8–11). There may also be a need to add new back-end systems to the existing network and/or modify the existing network to allow for more network traffic.

If you intend to add new back-end application functionality that does not currently exist, then you may need to design, develop, and deploy a web service and a set of business and data access objects that access the new back-end system. In this case, a data access object that allows information to be uploaded to or downloaded from the back-end system must be developed. Typically, an HTTP or SOAP interface is highly useful here.

8.3.7 Authentication

In a typical enterprise web application, users are allowed to browse the public areas of the web application without logging in. Anonymous users have access to public content, such as public documents, login, and enrollment pages, but a user must log in in order to access private, personalized, or secure content.

Figure 8–11 Existing back-end systems in an enterprise web architecture

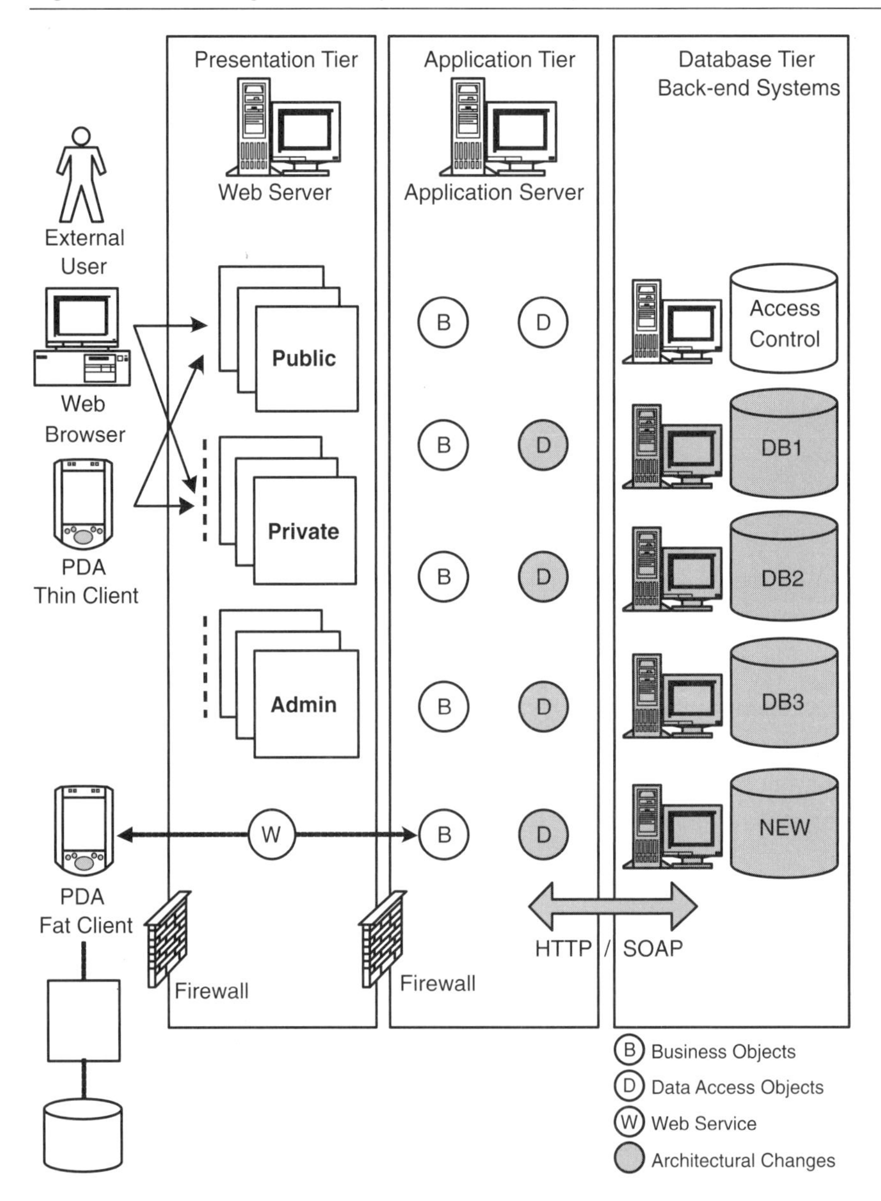

Authentication is provided in the form of a publicly accessible login page on the web application (see Figure 8–12). The user initially clicks on a login button or attempts to access a private page, object, URL, or resource without logging in. In these cases, the user is automatically redirected to the login page, where the user enters his/her username and password on a Web Form.

Upon Web Form submission, the login page passes the user's credentials to the authentication objects which validate the username and password against the access control database. Upon successful authentication, the authentication objects will issue an encrypted authentication token and any other tokens or cookies required (e.g., personal preferences). The user is then redirected to a private home page where he/she can access private, personalized or secure content. While the user is authenticated, access to the back-end systems is permitted through web pages and objects on the presentation and application servers.

Private pages on the presentation web server are protected by an authentication filter that prevents access to unauthorized users. The authentication filter checks for the authentication token to validate the session. A session is typically kept alive by user activity. When a user session expires, the user is redirected to the login page.

With the introduction of a population of external mobile users that may need to log in, the existing login functionality for external users will almost certainly have to be extended. Internal users may also have to be considered if the application services also apply to them.

If external users have continuous connectivity, then you will need to implement a set of small web pages that allow the users to log in using the same functionality as the enterprise web application. Therefore, you will need to implement the small page version of functionality that includes:

Login

- Authentication verification
- Authentication token(s) issuance
- Redirection to secured web pages
- Session establishment

Logout

- Authentication token(s) deletion
- Session closure

Ideally, you may be able to reuse the existing login authentication objects and only make changes to the login pages. However, if the authentication objects do not exist, you will either need to develop them or utilize the platform's authentication objects. In the case of Microsoft IIS, for example, you can utilize the "Forms Authentication" mechanism.

If the external mobile users are not always connected, you may need to implement a login form on the mobile device that checks login information in a local database to handle the contingency where the user is not connected to a server. When the mobile device is subsequently

Figure 8–12 Authentication in an enterprise web application

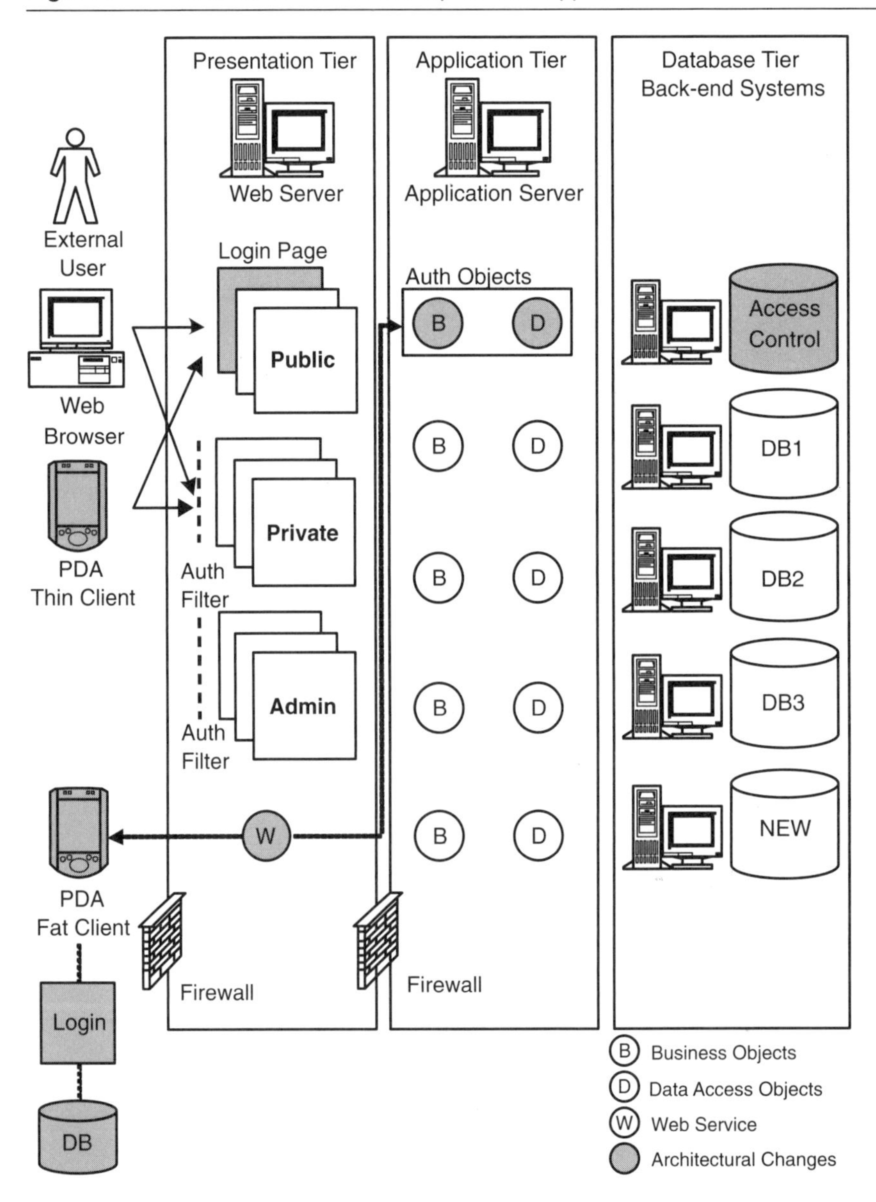

connected, the login information will have to be verified against the server. Integration with the existing authentication objects would be ideal. However, it is likely that a web service or some similar object that mediates between the mobile device and the authentication objects will need to be implemented.

It is interesting to note that with partially connected mobile devices, single sign-on for the enterprise as a whole becomes difficult because you may need to move part of the authentication authority mechanism to the mobile device and many enterprises are reluctant to place this measure of trust in mobile devices.

8.3.8 Enrollment

Large-scale enterprise web applications generally have an enrollment mechanism for all application services. Users are assigned a unique username based on the entry of the user's name, address, date of birth, mother's maiden name, social security number or national identity number, and other data available to uniquely identify the user. In addition, personal questions that only the user would typically know the answer to (e.g., first grade teacher's name) may also be requested. In a large-scale enterprise, this information is typically requested in the form of publicly accessible enrollment pages on the web application (see Figure 8–13).

With the introduction of a population of external mobile users that may need to enroll, the enrollment functionality for external users will almost certainly have to be extended. Therefore, you will need to implement the small page version of functionality that includes:

- User account creation
- User profile modification
- User deletion

If the external users will be always connected, then the implementation of a set of small web pages that allow the users to enroll will be necessary. Integration with the existing enrollment objects would be ideal.

If the users are not always connected, however, you may need to implement an enrollment page on the mobile device that temporarily stores enrollment information in a local database. When the mobile device is subsequently connected, the enrollment information is uploaded to the server. Integration with the existing enrollment objects would be ideal. However, it is likely that an object or a web service that mediates between the mobile device and the enrollment objects will need to be implemented.

Figure 8–13 Enrollment in an enterprise web application

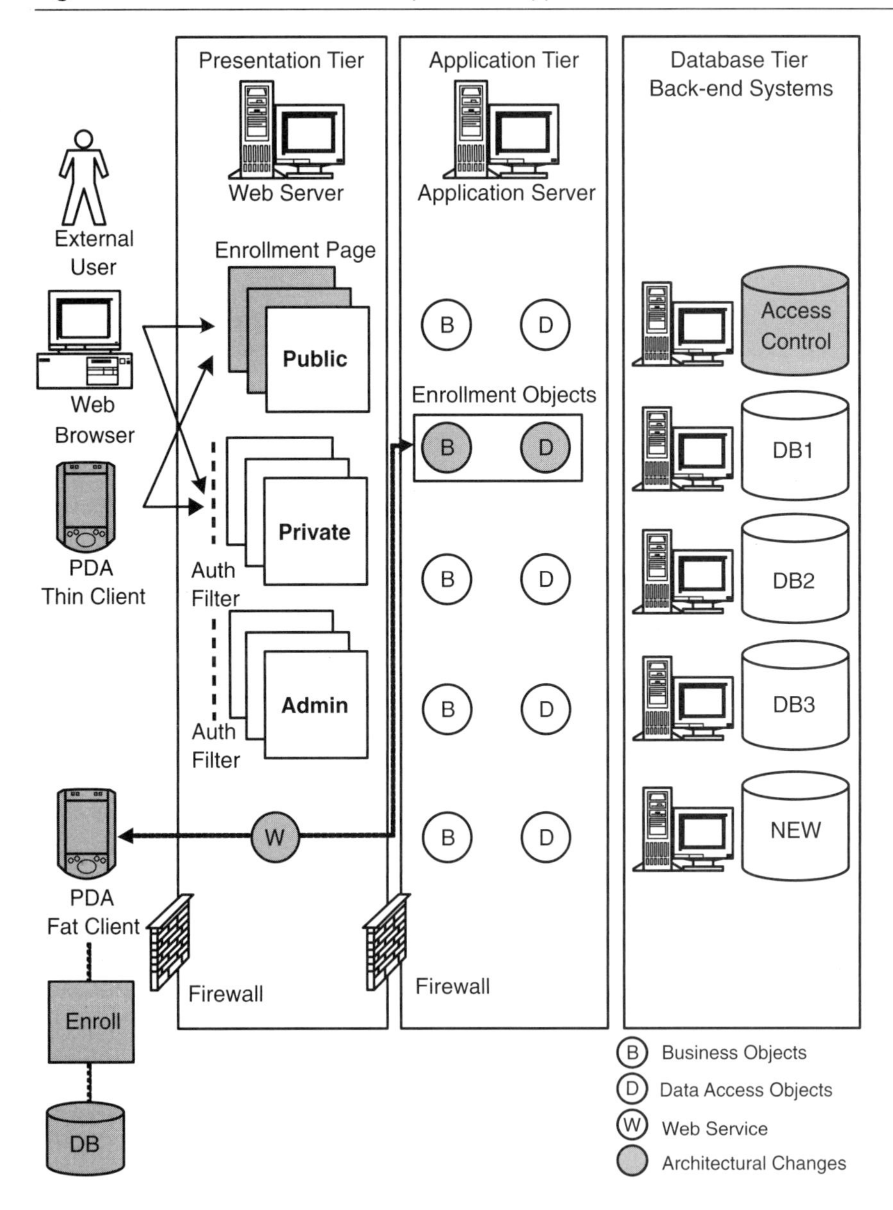

8.3.9 Administration

While most users will generally register through a self-service enrollment procedure, some human administrative support and intervention may nonetheless be needed to carry out functions such as user and group administration. In many enterprise web applications, these functions are typically provided in the form of an administration web application (see Figure 8–14).

With the introduction of a new mobile application, it is likely that the administrative tasks for internal users will have to be extended. External users typically don't use administration pages, so they can be effectively eliminated from consideration.

If your internal users will have continuous connectivity through their mobile devices, the implementation of a set of small web pages that allow administrators to perform mobile administration can be considered. These include:

User Administration

- Create a user account
- Modify a user account
- Delete or disable a user account

Role-Based Group Maintenance

- Create a group
- Modify a group
- Delete or disable a group
- Assign or unassign users to groups

Application Entitlements Maintenance

- Create application entitlements
- Modify application entitlements
- Delete or disable application entitlements
- Assign or unassign users or groups to application entitlements

Impersonation

- Perform any task a user can perform
- View any application a user can see

Web Application Management

- Start and stop applications
- Monitor applications
- Monitor performance
- Monitor security

Figure 8–14 Administration in an enterprise web application

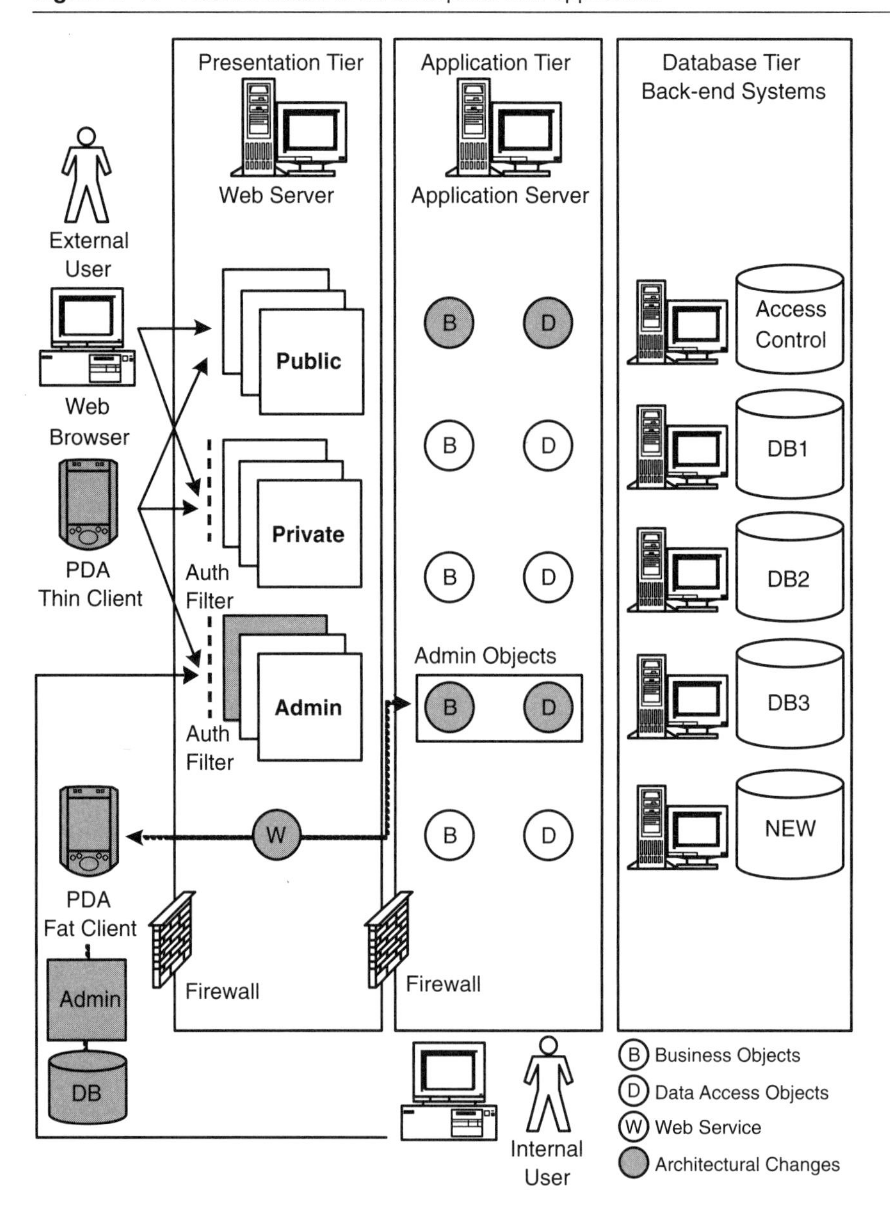

Perhaps the most difficult area to implement is the remote administration of an enterprise when administrators are not always connected using their mobile devices. Although perhaps not completely impossible, it is certainly very difficult for administrators to perform up-to-date monitoring of web applications or impersonate a user when they are not always connected. Thus, while certain tasks may be carried out while the administrator is disconnected, most must be done assuming continuous connectivity.

8.3.10 High Availability and Performance

The availability of an enterprise's applications and servers is of critical importance to the enterprise—a business application or server that is unavailable translates into money being lost. However, the more highly available the applications and servers are, the more expensive they are to develop and maintain. As a result, enterprise architectures vary considerably in their availability requirements.

Highly available servers and applications are typically able to fail over to another server when one becomes unavailable. This can be done in a variety of ways, including failing over to redundant systems and databases. Another way is to place a load balancer between the tiers (see Figure 8–15). When one server is down, the load balancer is able to redirect the traffic to another server. Thus, in addition to their normal traffic distribution functions, certain load balancers are also able to assist in failover.

Mobilizing the existing web architecture from a high availability perspective can be difficult because the front-end mobile applications and systems are still in their infancy and may not be able to cope with the demands made of them. Thus, if you have a highly available trading system, you may not get high availability if the front-end Presentation Tier and mobile clients are unable to cope.

The existing servers and applications may also have high performance parameters to handle large numbers of users or spikes in resource demand. For example, in an e-commerce web application, spikes might occur over the holiday season. In an e-business application, such as a trading web application, spikes might occur on a heavy trading day. The introduction of a new mobile application can have implications for the performance of the existing servers and applications. Due consideration must be given to the existing servers and applications if they are to perform at their normal level.

8.3.11 Scalability

Mobilizing an enterprise may mean that the number of devices and users your current enterprise architecture supports will increase. This can put a strain on your existing infrastructure.

While it is possible that the existing infrastructure is sufficient to handle the increase in mobile devices and user base, you may nonetheless need to scale and extend it to accommodate these possibly large increases in mobile devices, users, applications, and functions. Ideally, the

Figure 8–15 High performance and availability in an enterprise web architecture

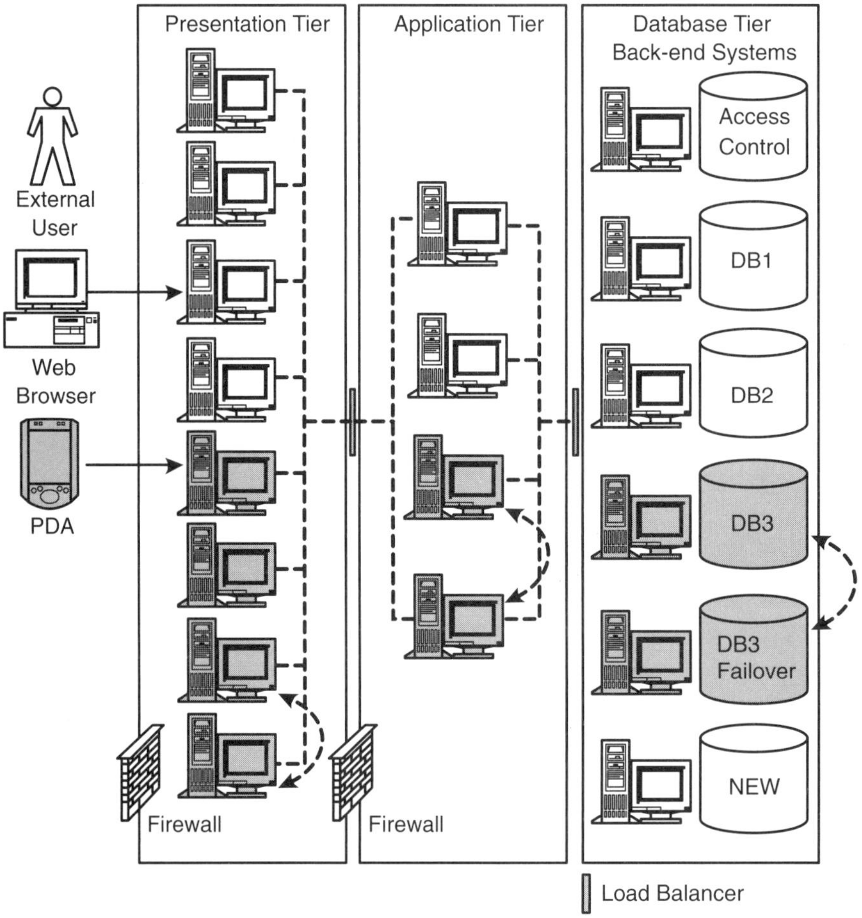

existing enterprise architecture has already been implemented so that both horizontal (adding more servers) and vertical (adding more powerful servers) scaling is possible without adversely affecting the existing applications.

A three-tiered architecture such as that illustrated in Figure 8–16 is scalable. For example, with the addition of new users on mobile devices, additional presentation and application servers or more powerful presentation and application servers can be added as required. Load balancers can be placed between any or all of the tiers to help distribute the traffic among the different servers. A code release of web pages to the Presentation Tier and objects to the Application Tier will allow the new servers to operate in an identical fashion to those already deployed.

Figure 8–16 Scalability in an enterprise web architecture

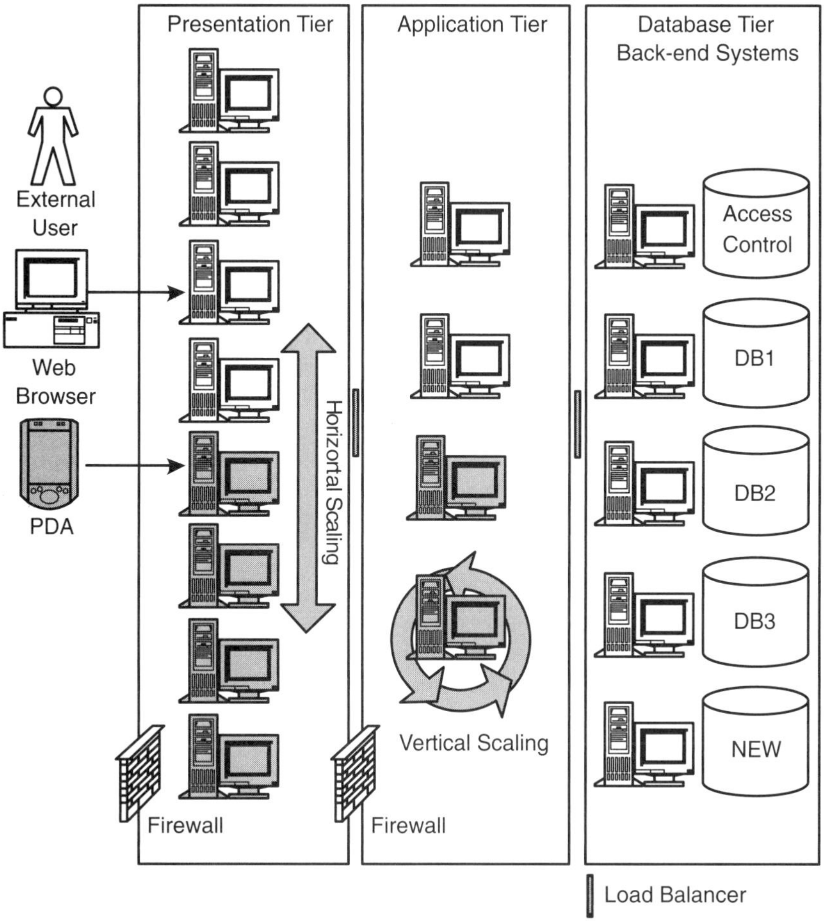

However, if your existing enterprise architecture is not scalable, mobilizing it may be very difficult. In this case, we suggest that you first modify it so that it is scalable before attempting to add mobile components.

8.3.12 Security

Introducing mobile clients that interface to existing enterprise web applications can result in security breaches within the system. As a result, it is important that any mobile access to the enterprise applications be strictly controlled and regulated. The last thing you want to do when you introduce mobile devices into your existing enterprise architecture is to breach your existing security. We will return to this important topic in Chapter 9.

8.4 SUMMARY

While existing enterprise web applications are not necessarily part of mobile application development, it is vital to consider them as part of the entire mobile application development process. Too often, the existing enterprise architecture is underplayed or ignored to the detriment of the mobile application and, possibly, to the undertaking as a whole.

We have attempted to share some of the major problems that are typically encountered when a mobile application is introduced into an existing enterprise architecture as well as some of the considerations that need to be taken into account in order to alleviate them.

Mobilizing existing applications is not easy even with world-class products and technologies. It requires application development, product integration, customization, infrastructure extensions, and people who can solve problems. Our experience is that there is no magic bullet and this may be the area that is the most difficult and expensive to undertake. However, the process can be made easier if due consideration is given to the architecture of your existing applications and infrastructure.

CHAPTER 9

Security

Quandoque bonus dormitat Homerus.
Even good Homer nods sometimes.
(Even the cleverest can make mistakes.)

—Horace

In this chapter, we describe the vulnerabilities and threats to a mobilized enterprise web architecture. We also describe mobile device security, secure communications between the client and servers, and security on the enterprise web servers and back-end systems.

9.1 MOBILIZED ENTERPRISE WEB ARCHITECTURES

At the enterprise level, a secure architecture is usually required to protect the service provider's assets from being attacked. Security architectures generally use some form of role-based access control and database-enabled or system-enabled authentication and authorization mechanisms that integrate user identity, user roles, and application entitlements. The security architecture may also provide additional services, such as encryption, certificate management, and smart card technology. In the following sections, we describe several of the most common vulnerabilities and threats.

9.1.1 Vulnerabilities

Mobilized enterprise web architectures are vulnerable to attack from many perspectives (see Figure 9–1). Some of these vulnerabilities are common and relatively easy to remedy. Others are more subtle and may not be initially intuitive or obvious. Some of the most common vulnerabilities are described in the following sections.

1. **User activity.** Users themselves can make the enterprise vulnerable. For example, they can inadvertently trigger an attack simply by opening an email attachment, downloading unsafe content from the Web, or running a script that contains a virus. The more privileged a user, the more dangerous this possibility is. For example, an administrator who inadvertently introduces a virus into a system can potentially propagate it throughout the enterprise.

2. **Weak usernames and passwords.** Weak usernames and passwords can also make an enterprise vulnerable. Typical weak usernames include "administrator" or "admin," while weak passwords are blank, "password," or "1234." Weak usernames and passwords can be remedied by instigating a security policy that enforces strong usernames and passwords. For example, you can require that a password must be at least ten characters in length, contain uppercase and lowercase alphabetic and numeric characters, and include at least one punctuation mark.
3. **Excessive permissions.** Users are often granted more permissions and privileges than are strictly necessary. This may allow malicious users to gain greater access to the system's resources than should be possible. It also allows users to accidentally commit security breaches.
4. **Deception.** Users can be deceived into revealing private information about themselves. For example, an attacker posing as an online help desk operator might persuade a user to reveal his/her password under the pretext of performing some administrative task. A well-written application should never expose a password or require a user to expose his/her password.
5. **Excessive services and ports.** Certain web servers (e.g., Microsoft IIS) enable more services and port connections by default than are strictly necessary to run a specific application. These unused services and ports can provide an opening for attackers.
6. **Unencrypted data transfer.** Data is typically sent between a client and server in unencrypted form. If this data is intercepted on the wire or wirelessly, an attacker can read and/or alter the data.
7. **Buffer overrun.** Buffer overrun attacks occur when malicious users exploit an unchecked buffer in a program and overwrite the contents with their own code or data. If program code is overwritten with new code, this changes the program's operation to that prescribed by the attacker. If data is overwritten with other data, this corrupts the data's integrity.
8. **SQL injection attacks.** SQL injection attacks can occur when SQL statements are dynamically built through user input. Attackers with good SQL knowledge can modify the SQL statements and make the application execute tasks that may expose passwords or other information.
9. **Code-embedded usernames and passwords.** Attackers can gain access to a system if they are able to read application code that contains embedded usernames or passwords. Web pages that contain database connection strings are vulnerable in this scenario.

Figure 9–1 Mobilized enterprise web architecture vulnerabilities

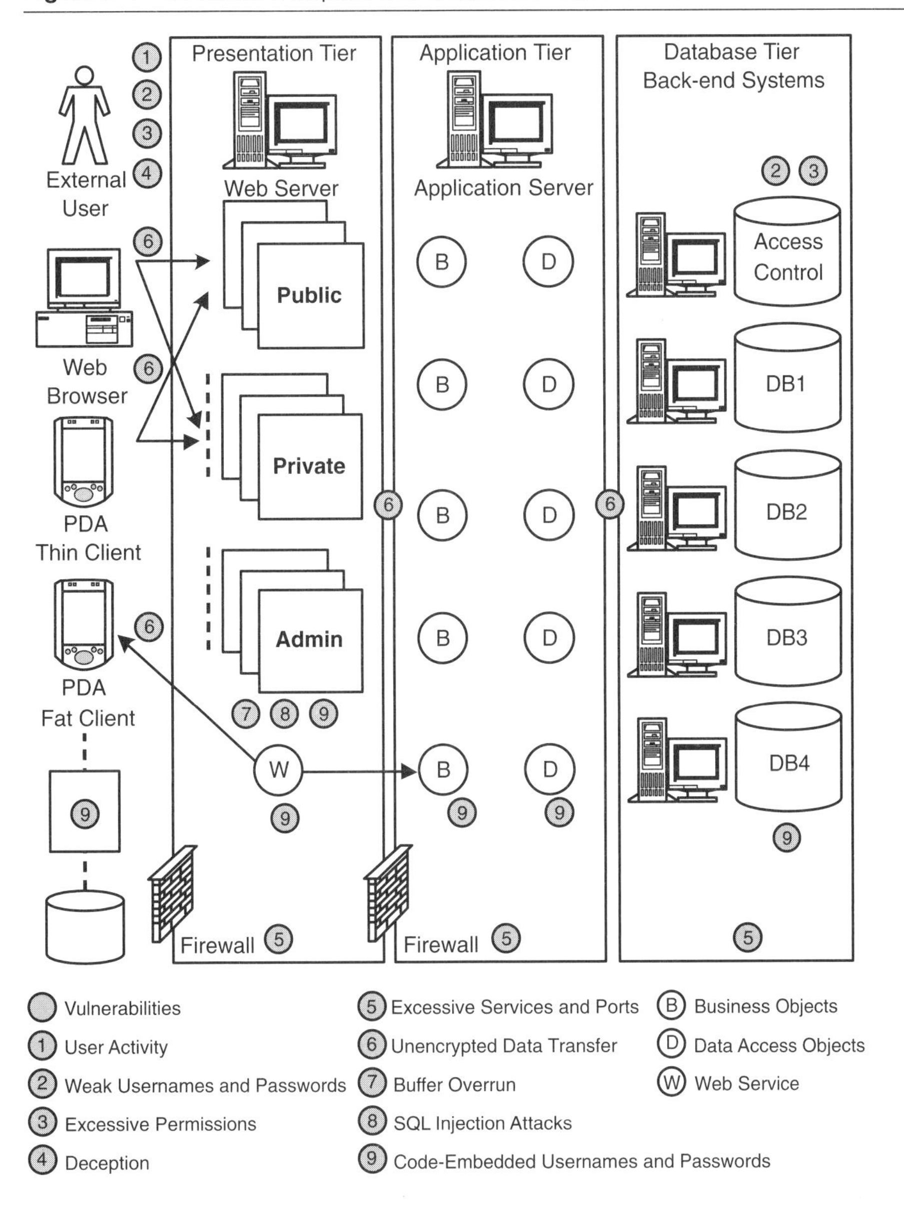

9.1.2 Threats

Many threats can be made against an enterprise's systems and applications. Microsoft classifies these threats into several major categories that are commonly known and easily remembered by the acronym "STRIDE." Each letter of STRIDE represents a type of threat: Spoofing, Tampering, Repudiation, Information disclosure, Denial of service, and Elevation of privilege.

1. **Spoofing.** Spoofing occurs when an attacker pretends to be a user who is trusted by the system. In doing so, the attacker is able to assume the identity of the user and may gain access to unauthorized areas. Spoofing can also occur when a system is fooled into trusting another system that is under the control of an attacker.
2. **Tampering.** Tampering occurs when an attacker gains access to a system and tampers with its operation or function.
3. **Repudiation.** Repudiation occurs when a system is unable to prove that an event occurred. This is normally a result of insufficient auditing or logging abilities. For example, even if an attacker is caught, he/she may be able to repudiate having attacked the system because there is no proof.
4. **Information disclosure.** Information disclosure occurs when an attacker is able to view an individual's private data. Some of this data obviously needs to be secure, such as a user's date of birth, mother's maiden name, or social security number. However, other information may also require the same level of security, but for more subtle reasons. For example, a user's selection of favorite stocks does not seem at first glance to require strong security. However, if the stock portfolio belongs to a prominent investor, it could be leveraged by an attacker.
5. **Denial of service.** A denial of service attack disrupts or prevents access to a system or its resources by flooding the system with messages. These attacks cause applications to become starved of CPU or memory resources and ultimately prevent users from accessing the applications.
6. **Elevation of privilege.** Elevation of privilege attacks are launched when an attacker is able to elevate or gain additional privileges to those normally granted (e.g., a normal registered user is able to gain administrative privileges).

9.1.3 Mitigation

Vulnerabilities and threats can be mitigated by a variety of mechanisms. Some of the most common mitigation methods are described below.

Authentication and access control. Authentication can be used to verify that a user is who he/she claims to be, while access control through the use of entitlements determines what the user is allowed to do and see.

Secure communications. Communications between the various tiers can be secured using encryption so that even if a message is intercepted, it cannot be deciphered.

Quality of service. A software service that profiles messages sent to the system can be used to mitigate denial of service attacks.

Throttling. Throttling limits the number of messages that can be sent to a system. If throttles are set, it is possible to mitigate denial of service attacks because the system cannot be flooded with messages.

Auditing. Logging important events (auditing) is an important part of mitigation; it creates a record of system and user activity that can be reviewed by system owners to determine security threats and their sources.

Filtering. Authentication filters can be set on protected resources to evaluate and block all messages coming into the enterprise that pose a threat.

Least privilege. Users should be granted the least number or level of privileges that will allow them to carry out their normal tasks.

The rest of this chapter describes various threats against the mobilized enterprise web architecture in detail and the mitigation techniques used to defend the enterprise against them.

9.2 USER-TO-MOBILE CLIENT SECURITY ISSUES

The simple interaction between a user and a mobile device can lead to security breaches. For example, some of the vulnerabilities in this area include user activity, weak usernames and passwords, excessive permissions, and users being deceived into revealing too much information.

There are several mechanisms available to alleviate these vulnerabilities, including the use of authentication, smart cards, and biometric authentication (see Figure 9–2). In addition, granting the minimum level of privileges each user needs will help alleviate these vulnerabilities. In the following sections, we will discuss these mechanisms in more detail.

9.2.1 Authentication

Authentication helps mitigate the threat of spoofing. If the application mandates the use of strong usernames and passwords and sets minimal required privileges, the enterprise application's vulnerability to attack will be reduced. Authentication can be carried out at several levels, as described below:

1. **Mobile device.** The mobile device itself may have or require the entry of a username or password after a period of time has elapsed. For example, many cellular telephones require the entry of a password before they will call a number.
2. **Application authentication (local).** If the mobile client is a fat client, the application may have a login page that authenticates against a local database when the mobile application is disconnected from the back-end servers. A user must log in to use a particular application regardless of whether he/she is connected to the enterprise servers. This means that additional security, such as an encrypted local database, may be required on the mobile device.

Figure 9–2 User-to-mobile client security issues

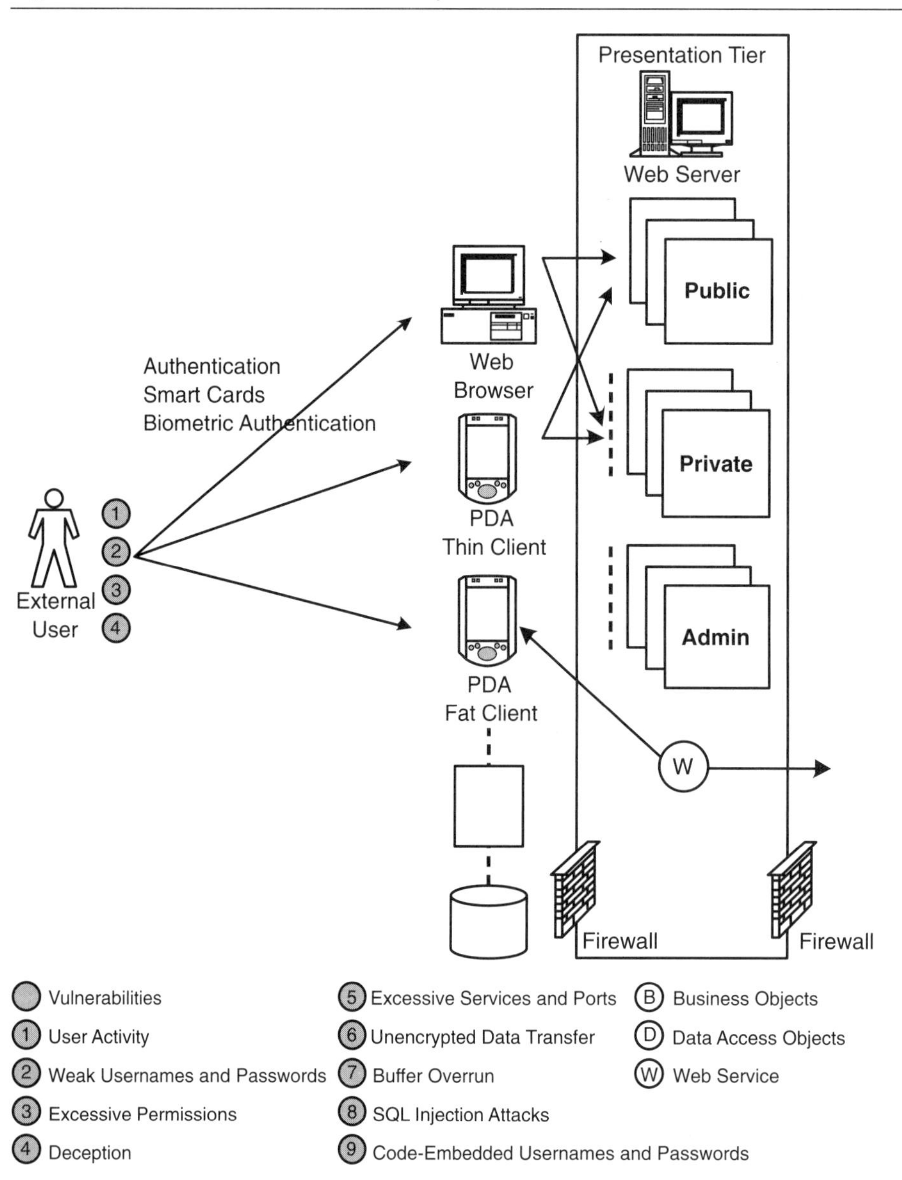

3. **Application authentication (remote).** If the mobile client is a thin client, no custom application pages reside on it. A user can still log in to the application on the server, but the user must then be always connected. Authentication is therefore remote and is handled by security on the server. However, you should ensure that persistent state cookies, automatic password fill-in capabilities, and other such mechanisms are not a threat to the enterprise as a whole. Any application that requires the storage of this type of information on the client can be vulnerable to attack since the person who next uses the device may not be the authorized user.

9.2.2 Smart Cards

The practice of implementing smart cards to validate a user's identity is now also in use, but it is not yet as widespread as simple logging in and logging out.

9.2.3 Biometric Authentication

Emerging technologies that involve biometric authentication such as fingerprinting, voice recognition, and retinal scans are also becoming viable options for establishing user identity. Biometric authentication helps mitigate the threat of spoofing and drastically reduces the enterprise application's vulnerability to attack, since the identity of a user is known very precisely, reducing the chance of misidentification to virtually zero.

9.3 MOBILE CLIENT SECURITY ISSUES

Users are understandably concerned about the security and privacy of data on mobile devices, especially since mobile devices tend to be easily misplaced, lost, or stolen.

However, while the loss of a mobile device may be annoying and expensive for the user, the loss of the information stored on the mobile device may far outweigh the value of the mobile device itself. For example, the loss of a mobile device containing records of orders taken but not yet uploaded to the server may be very costly. In addition, the possible exposure of private information stored on the mobile device to unauthorized persons (i.e., the person who stole the device or some other intruder) may have serious consequences.

As a result of these concerns, several mechanisms can be implemented to better protect mobile devices including automatic logout, credentials re-entry, data destruction, database encryption, and encryption of code-embedded usernames and passwords (see Figure 9–3). The following sections discuss these mechanisms in more detail. Note, however, that these mechanisms are not necessarily foolproof. In practice, it is very difficult to completely guarantee the safeguarding of mobile devices. Implementing these safeguards, however, makes it very difficult for intruders to access secure data even if they physically have the mobile device.

Figure 9–3 Mobile client security issues

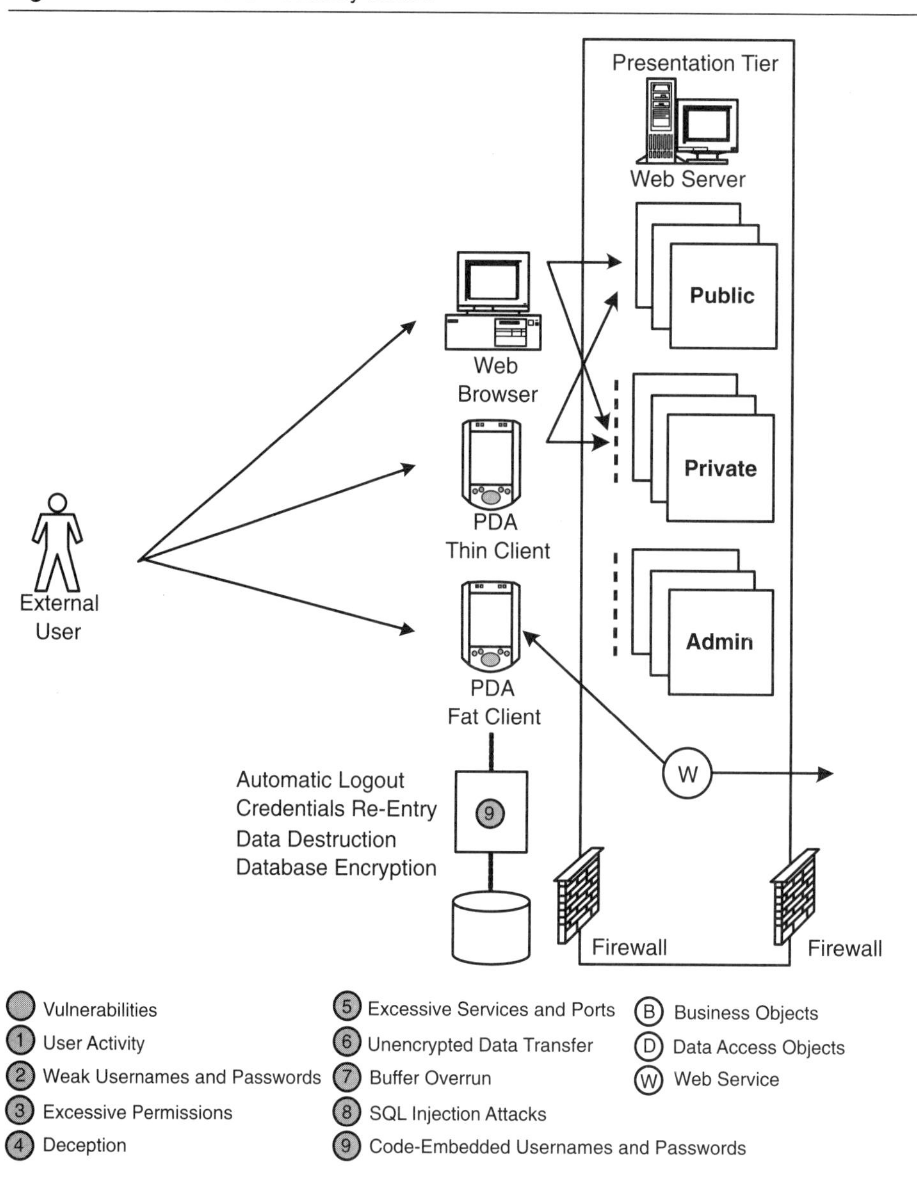

9.3.1 Automatic Logout and Credentials Re-Entry

Many mobile devices have automatic logout features triggered by user inactivity. These features serve to protect the data on the mobile device if a user loses or has the device stolen.

For example, cellular telephones, Pocket PCs, and Tablet PCs can be set to require re-entry of user credentials after some period of time has elapsed. The necessary software is provided by the vendor and it is typically a built-in feature of the product.

9.3.2 Data Destruction

It is possible to implement a custom data destruction mechanism that erases sensitive programs and data from the mobile device if it is out of contact from the server for a prolonged period of time.

The mechanism can be a simple program that deletes designated files or data from the mobile device after a predefined period of time has elapsed. It can also be triggered to run when a user enters an incorrect password a certain number of times.

It is also possible to create a server-initiated program that deletes designated files from the mobile device. This may be harder to implement, however, since you may not be able to contact the mobile device to initiate such an action.

9.3.3 Database Encryption

It is possible to encrypt the data in certain mobile device databases. For example, Microsoft SQL Server CE databases can be encrypted. Thus, even if a Pocket PC is lost, it becomes extremely difficult for intruders to read the data held within the database.

9.3.4 Code-Embedded Usernames and Passwords

Encrypting the mobile device's database, however, will not help if an attacker can gain access to application code that contains embedded usernames or passwords. As a result, all usernames and passwords contained in the mobile client code or in configuration files should also be encrypted.

9.4 CLIENT-SERVER COMMUNICATIONS SECURITY ISSUES

Internet communications are not inherently secure, and it is possible to intercept transmissions between a client and server. As a result, several types of communication encryption have been developed to help prevent tapping into the message stream and intercepting messages (see Figure 9–4).

9.4.1 Communication Encryption

The use of encryption is perhaps the most common method used to prevent the reading of clear text messages between a client and server.

For example, the use of Secure Sockets Layer (SSL) is one of the most common mechanisms used to prevent clear text message interception over the Internet. SSL is typically used to encrypt data between a client and a web presentation server.

Figure 9–4 Client-server communications security issues

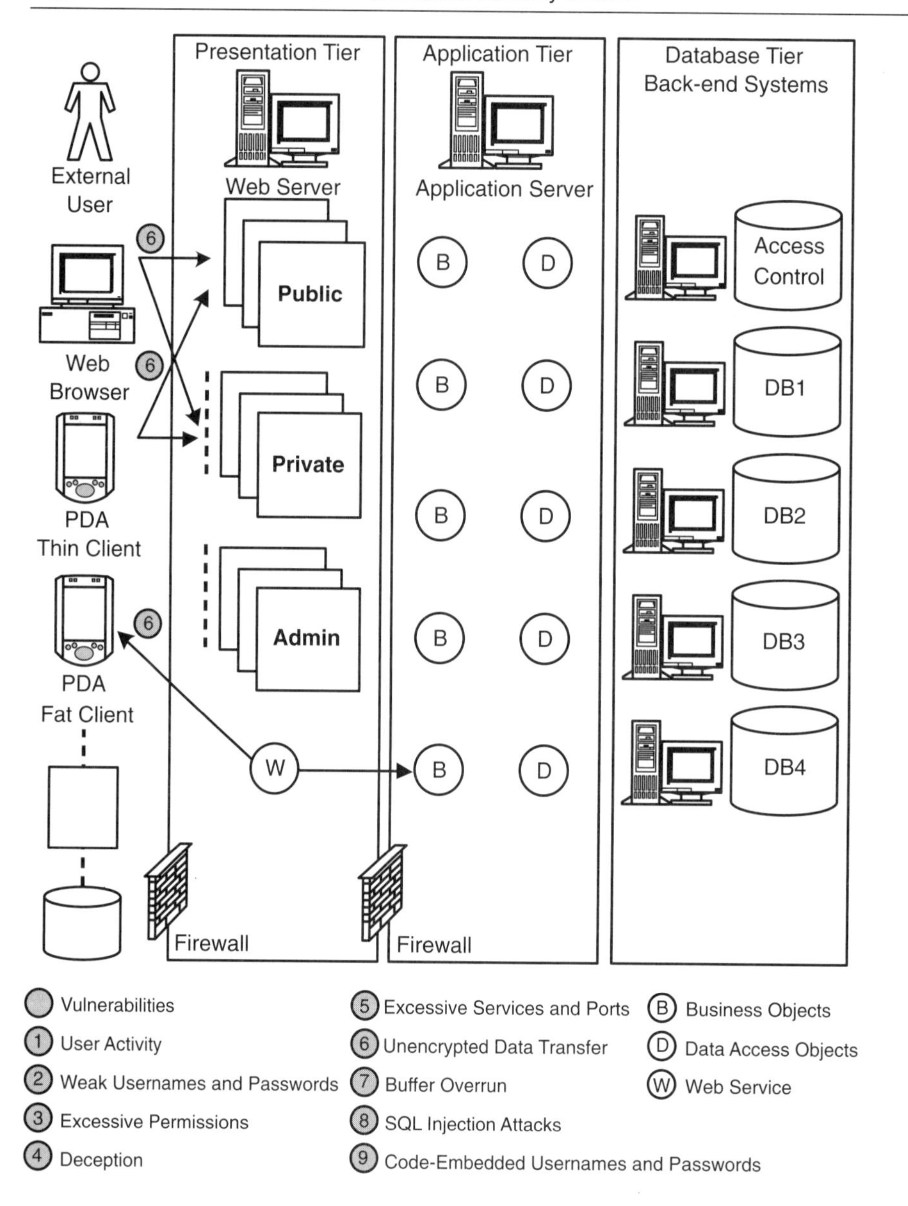

- Vulnerabilities
- (1) User Activity
- (2) Weak Usernames and Passwords
- (3) Excessive Permissions
- (4) Deception
- (5) Excessive Services and Ports
- (6) Unencrypted Data Transfer
- (7) Buffer Overrun
- (8) SQL Injection Attacks
- (9) Code-Embedded Usernames and Passwords
- (B) Business Objects
- (D) Data Access Objects
- (W) Web Service

In addition, wireless connections are often encrypted using Wired Equivalent Privacy (WEP). The use of such an encrypted wireless connection requires the entry of an encryption key on the mobile device. Without the encryption, it is possible for unauthorized intruders to utilize or intercept information from the wireless network.

While the use of encryption is generally recommended, it is also pertinent to note that encryption comes at a price. Encryption (and decryption) can be quite resource-intensive for both client and server. As a result, it is often better to selectively encrypt data than to encrypt everything.

9.5 EXISTING WEB ARCHITECTURES AND BACK-END SYSTEMS SECURITY ISSUES

The existing web architecture and back-end systems may also have certain vulnerabilities (see Figure 9–5). In the following sections, we discuss some of the standard mechanisms that are used to secure enterprise web architectures and back-end systems, such as firewalls, port lockdown, communication encryption, and database authentication and encryption.

9.5.1 Firewalls and Tier Separation

The judicious placement of firewalls and the separation of code and logic into tiers are just two of several powerful mechanisms used to secure enterprise web architectures.

Firewalls can be set up to filter all network traffic moving in and out of the enterprise. Highly secure architectures may utilize multiple layers of firewall protection to create several regions of trust. The frontmost firewall guards the enterprise domain from the Internet, while the innermost firewall defends the corporate intranet. Again, the semitrusted region between these two firewalls is typically called the DMZ. This area may be hosted outside of the enterprise by an Internet service provider.

The use of two firewalls creates three zones: the public, untrusted Internet zone; the semi-trusted DMZ; and the trusted intranet zone. One benefit to using a three-tier application architecture, from a security perspective, is the ability to distribute application components securely within each zone. For example, many enterprises place the Presentation Tier in the DMZ and the Application and Database Tiers within the corporate intranet. This design provides multiple layers of security protecting the crucial data stored within the database.

9.5.2 Application Services and Port Lockdown

As mentioned earlier in this chapter, more application services and ports may be available than are actually required to operate the application. Generally, prior to deployment, a review of all services and open ports should be undertaken, and unnecessary services stopped or disabled and unused ports closed or disabled.

Figure 9–5 Existing web architecture and back-end systems security issues

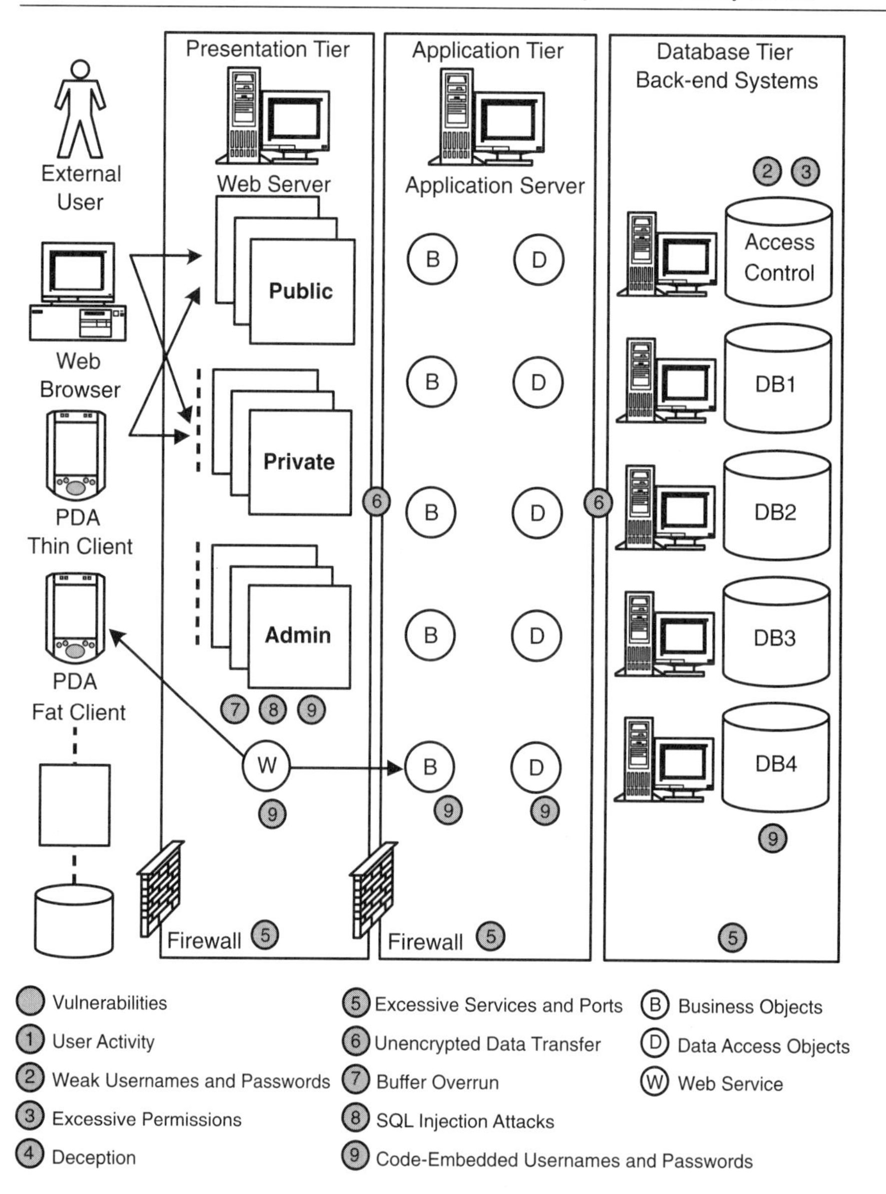

9.5.3 Communication Encryption

Although the application, database, and back-end servers typically lie within an enterprise's intranet, certain highly secured intranets (e.g., within banks or brokerages) may require additional encryption on the network between them. This is to prevent unauthorized access by internal users.

There are several commonly used encryption mechanisms to prevent the reading of clear text messages between servers, as described below.

SSL. SSL is one of the most common mechanisms used to prevent clear text message interception within an enterprise. SSL may be used to encrypt data between the application and database servers.

IPSec. Internet Protocol Security (IPSec) is also used to encrypt data passing between back-end systems, such as the web server, application server, and database server.

RPC encryption. The use of Remote Procedure Call (RPC) encryption is also very common. It is typically used to encrypt messages between serviced components on remote servers.

9.5.4 Database Authentication

In enterprise applications, database authentication can be performed using individual user accounts or service accounts. Each method has its advantages and disadvantages. Individual user authentication requires maintenance of multiple accounts within the database and minimizes the ability for applications to pool database connections. The use of a service account, however, creates a different problem. If an attacker is able to retrieve the service account username and password, he/she may gain access to the entire database. In order to prevent this, applications that use service accounts should keep the database connection string in an encrypted file for safekeeping.

9.5.5 Database Encryption

Data held within the enterprise databases and back-end systems may be stored encrypted. While the database can generally be expected to lie within an enterprise's intranet, certain highly secured intranets (e.g., within banks or brokerages) may require that the database be encrypted to prevent unauthorized access by internal users.

9.6 SUMMARY

Security in a mobilized enterprise web architecture is an umbrella concern that encompasses users, mobile devices, mobile clients, networks, web applications, and back-end applications.

Every system and interface can be subjected to some form of attack and needs to be secured. However, a secure architecture is only as strong as its weakest link; once one system has been breached, the entire enterprise can be compromised.

It is also important to note that no single security feature is completely foolproof. Generally, several security features should be used in combination to reduce the threat to the minimum level acceptable to the enterprise.

Security is a very broad topic and a detailed treatment of the subject is well beyond the scope of this book. We recommend that interested readers research the concepts we have addressed in more detail through the literature or work of experts in the field.

CHAPTER 10

Mobile Application Development Management

I will knog your urinals about your knave's cogscomb for missing your meetings & appointments.

—Shakespeare

This chapter discusses the management of mobile application development projects. We start by describing some of the normal activities in project management, followed by specific activities that need to be performed when managing mobile application development projects, including gathering requirements, designing, coding, testing, deploying, and operating the application.

10.1 PROJECT MANAGEMENT

In many respects, the management of mobile application development projects is similar to the management of any major enterprise application development project. Enterprise application development projects are typically broken down into phases, during which activities such as gathering requirements and designing, coding, testing, and deploying the application can be performed. These phases can be arranged differently depending on the particular management model that is employed. Project management also consists of other activities, including project planning, resource management, gathering and disseminating project status, and working on project financials. The following sections describe the above activities in more detail.

10.1.1 Management Models

Most application development projects consist of similar phases, during which sets of activities are completed. These phases typically include:

- Requirements Phase—Gathering requirements
- Design Phase—Designing the application

- Coding Phase—Coding and integrating the components
- System Test Phase—Integration and system testing
- Deployment Phase—Deploying and releasing the application

While there are differences in what activities are performed and what is produced in each phase, the major difference between management models lies in how the various phases within a project are organized.

Two popular management models that use imaginative ways of organizing time, people, and resources to complete activities are the waterfall model and the iterative and incremental spirals model.

1. **Waterfall.** Many classic application development projects followed the waterfall process (see Figure 10–1). In this process, projects proceed from the Requirements phase through the Design, Coding, and System Testing phases before finally reaching the Deployment phase.

   The problem with the waterfall model is that each phase relies on the preceding phase being complete before the project can continue. Thus, requirements have to be complete before the design process can begin, and the design process has to be complete before coding can begin.

   In practice, this makes the waterfall model somewhat inflexible and not very adaptable, especially when complex or innovative projects are undertaken. Requirements do change and so, correspondingly, does the design and code.

   It also gets progressively more difficult to correct or change an item once it is completed. For example, design problems may not be discovered until very late in the project when it is extremely risky and costly to change them.

Figure 10–1 Waterfall model

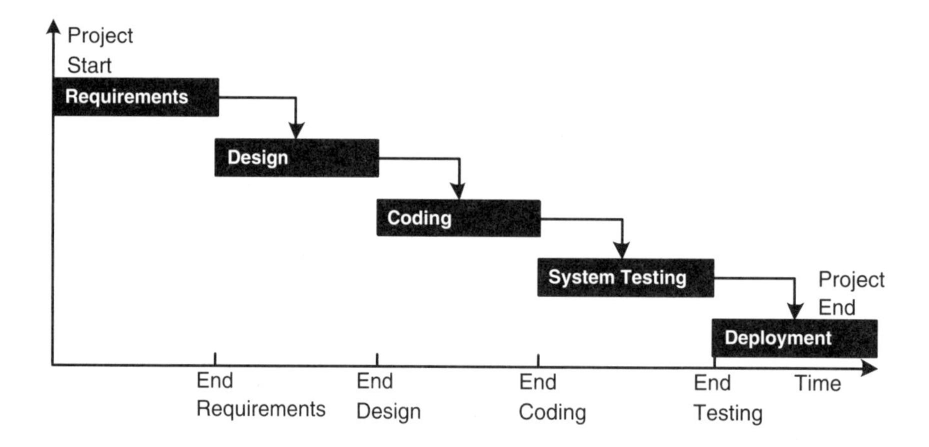

2. **Iterative and Incremental Spirals.** More modern approaches to project management, such as the Rational Unified Process (RUP), are based on a variation of Boehm's spiral model, which recommends a more iterative and incremental approach to the development process (see Figure 10–2).

   In this model, you initially plan, gather requirements, design, code, test, and deploy the application rapidly. You then re-evaluate your results and carry out this procedure again (iteratively). Each release gets progressively (incrementally) better in terms of documentation and functionality until the application is finally deemed acceptable. The idea is that with each release, you get closer to your end goal until you finally reach it. Essentially, you divide and conquer by taking numerous steps towards your goal while adjusting your route along the way.

   There are several benefits to using this approach. First, the risks associated with the project diminish over time. Second, the project tangibly progresses and concrete evidence of improvement is seen with each passing release. Delivering an operational application early also helps clarify poorly formulated requirements and highlights problems that need to be addressed. The first release can also be treated as a prototype, in which people are able to see a working model of the final application early in the process.

Figure 10–2 Iterative and incremental spirals model

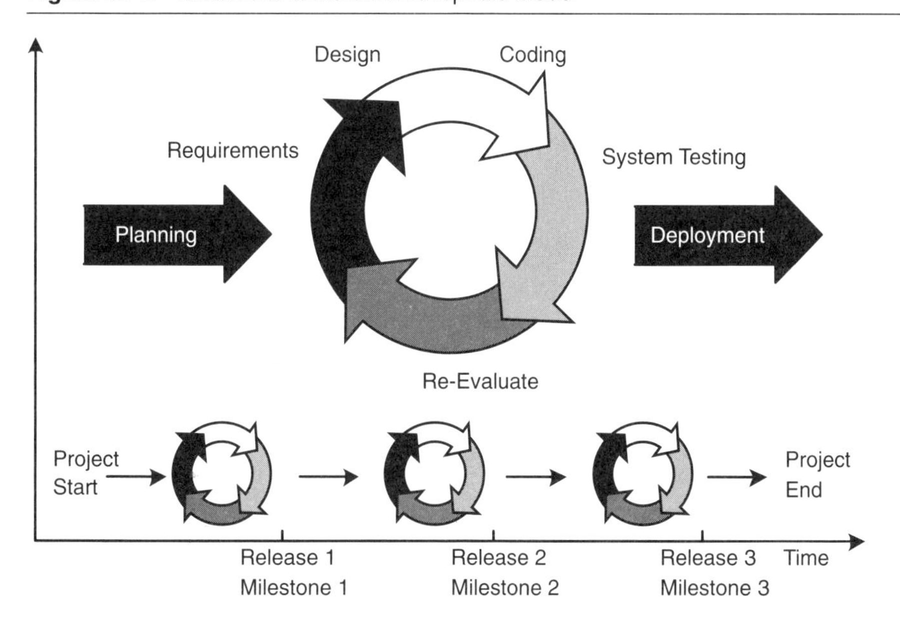

10.1.2 Planning

Planning is the organization of limited resources to address a set of tasks in order to achieve your goal. Resources that are typically limited include:

- People (the team)
- Time
- Materials (hardware, software, tools, SDKs)
- Money

A project plan is often of great help in organizing a project. Essentially, a project plan contains all the tasks and subtasks that have to be carried out in order to achieve your goal. These tasks are organized by time, and typically include task-specific completion dates, along with the names of the team members that are assigned to carrying out the tasks.

In the simple project plan illustrated in Figure 10–3, for example, we assume that three people are building a fat client mobile application that is normally disconnected from a web site. When the mobile device is connected, the web site must be able to accept and store the information gathered from the mobile device. Communication between the mobile client and web site is handled by a web service.

The project spans two releases. The team plans, gathers requirements, designs, codes, tests, and deploys the first release. The team then evaluates the first release and goes through the cycle again before deploying the second and final release.

In an enterprise application development project, there can be a large number of tasks and subtasks, with complex interdependencies and relationships between them. However, there are many excellent tools, such as Microsoft Project, to help you plan the project.

It is important that you plan ahead and then stick to your plan as well as you can. However, you should have enough flexibility to change the plan if you find things are not working out. It is also important to move on when time is up for a certain scheduled task. It is very tempting to keep going for just a little bit longer just to get something a little bit better. If you do so, however, upcoming items may be affected.

10.1.3 Team Resources

Development teams are generally composed of a mix of the following personnel:

- Coders or specialists (e.g., graphic designers, industry experts, etc.)
- Technical leader
- Project manager

Figure 10–3 Project plan

	Task Name	Duration	Start	Finish	Resource
1	Project Start	70 days	Mon 8/4/03	Fri 11/7/03	VL
2	**⊟ Planning 1.0**	**15 days**	**Mon 8/4/03**	**Fri 8/22/03**	**VL**
3	Project Plan	1 day	Mon 8/4/03	Mon 8/4/03	VL
4	Requirements	7 days	Tue 8/5/03	Wed 8/13/03	VL
5	Design	7 days	Thu 8/14/03	Fri 8/22/03	VL
6	**⊟ Implementation 1.0**	**30 days**	**Mon 8/25/03**	**Fri 10/3/03**	**VL**
7	Mobile Client	30 days	Mon 8/25/03	Fri 10/3/03	HS
8	Web Service	30 days	Mon 8/25/03	Fri 10/3/03	RS
9	Web Site	30 days	Mon 8/25/03	Fri 10/3/03	VL
10	**⊟ Testing 1.0**	**10 days**	**Mon 10/6/03**	**Fri 10/17/03**	**VL**
11	Mobile Client	10 days	Mon 10/6/03	Fri 10/17/03	HS
12	Web Service	10 days	Mon 10/6/03	Fri 10/17/03	RS
13	Web Site	5 days	Mon 10/6/03	Fri 10/10/03	VL
14	**⊟ Deployment 1.0**	**1 day**	**Mon 10/13/03**	**Mon 10/13/03**	**VL**
15	Release 1.0	1 day	Mon 10/13/03	Mon 10/13/03	VL
16	**⊟ Planning 2.0**	**4 days**	**Tue 10/14/03**	**Fri 10/17/03**	**VL**
17	Project Plan	1 day	Tue 10/14/03	Tue 10/14/03	VL
18	Requirements	1 day	Wed 10/15/03	Wed 10/15/03	VL
19	Design	2 days	Thu 10/16/03	Fri 10/17/03	VL
20	**⊟ Implementation 2.0**	**10 days**	**Mon 10/20/03**	**Fri 10/31/03**	**VL**
21	Mobile Client	10 days	Mon 10/20/03	Fri 10/31/03	HS
22	Web Service	10 days	Mon 10/20/03	Fri 10/31/03	RS
23	Web Site	10 days	Mon 10/20/03	Fri 10/31/03	VL
24	**⊟ Testing 2.0**	**4 days**	**Mon 11/3/03**	**Thu 11/6/03**	**VL**
25	Mobile Client	4 days	Mon 11/3/03	Thu 11/6/03	HS
26	Web Service	4 days	Mon 11/3/03	Thu 11/6/03	RS
27	Web Site	4 days	Mon 11/3/03	Thu 11/6/03	VL
28	**⊟ Deployment 2.0**	**1 day**	**Wed 11/5/03**	**Wed 11/5/03**	**VL**
29	Release 2.0	1 day	Wed 11/5/03	Wed 11/5/03	VL
30	Project End	0 days	Fri 11/7/03	Fri 11/7/03	VL

Code developers typically have a mix of experience and skills with mobile and web development. The specific language and technology skill requirements for coders depend somewhat on the specific project, but a good combination may include knowledge of HTML, XML/XSL, VB.NET, C#, and Java, with technology expertise in .NET, J2ME, and J2EE.

The technical leader normally has a coding background and is a senior-level developer. He/she also generally has managerial, documentation, architecture, and coding skills. The project manager also needs managerial and documentation skills. The project manager does not necessarily require technical expertise, although it can facilitate the process if he/she does.

For small projects with one to five developers, we generally suggest that the development team be composed of a technical leader and several coders. If there is no project manager, the technical leader typically assumes the tasks a project manager performs in addition to those normally assigned.

For medium to large development projects with six to twenty developers, we generally suggest small teams of highly trained personnel. For example, there may be several technical leaders, each responsible for significant areas of functionality. Several coders would typically report to each technical leader and each technical leader would report to the project manager. For example, in a large application development effort, you might have one project manager, three technical leaders (e.g., one responsible for the mobile application, one responsible for the web services, and one responsible for the web site), and several coders reporting to each technical leader.

In addition to the development team, an enterprise often has other teams that both affect the project and are affected by it (see Figure 10–4). For example, the project manager may directly report within the IT Development department of an enterprise, but he/she might also indirectly report to or confer with a variety of people in different teams. Descriptions of some of these other teams are provided below.

Business. Business personnel carry out functions that are critical to the organization from a business perspective. For example, in a bank, these employees might be financial advisors. If the application affects the way the enterprise carries out its business, the business personnel must be included in project discussions.

Sales. Sales personnel typically sell the products or services of the enterprise. If the application affects the way the enterprise sells its products or services, it is important to involve sales personnel.

Testing. In large enterprises, there may be a separate group that handles the system testing of enterprise applications. These employees test code releases from the developers and may carry out releases to production.

Operations. Operations staff may include operators, help desk personnel, and administrators. Operators ensure that the enterprise application remains operational. They also handle operational aspects of the application (e.g., manage code and document releases from development, perform the release, etc.). Help desk personnel provide assistance to all users of the enterprise application. System administrators have the highest privileges and may add, modify, and/or delete help desk personnel, operators, and other users. They may also perform other privileged system administration tasks.

Figure 10–4 Development project team

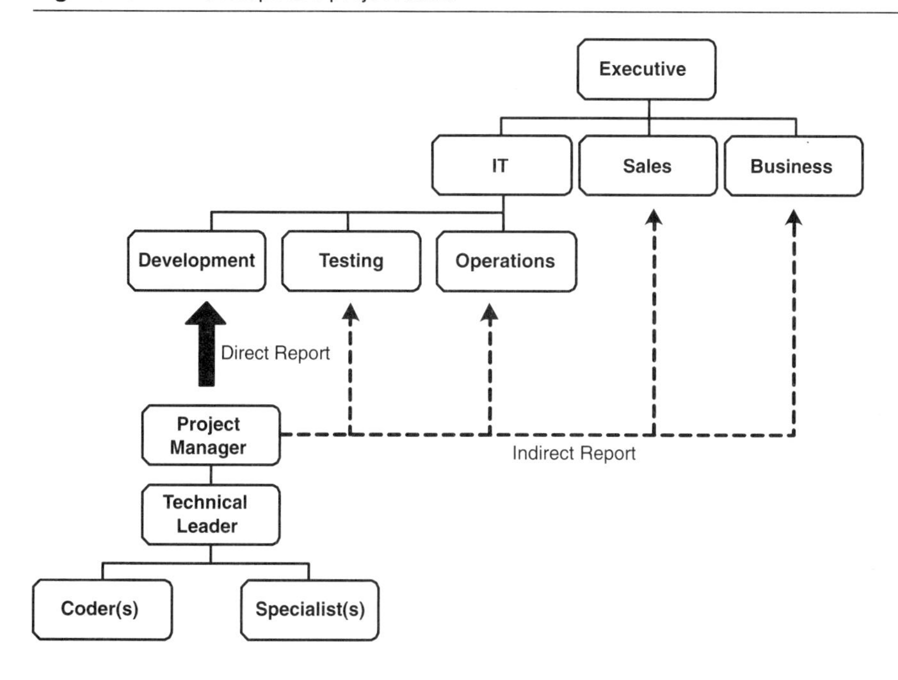

10.1.4 Status Reports and Status Meetings

In order to help facilitate communication of project status between all relevant parties, we suggest that each member of the project development team produces a weekly status report. The status report would typically have a simple structure and consist of four sections, as follows:

1. Completed Tasks—Tasks done and dates completed
2. Open Tasks—Tasks to do and dates to be completed
3. Issues/Problems—Issues and problems
4. Time Spent—Time spent on tasks for the week

Each team member fills in his/her status report and emails it to the project manager. The project manager consolidates them with his/her own status report, which is the same as each team member's report with one additional section that summarizes the project status (see Figure 10–5). The report should be named status_yyyymmdd.doc where yyyymmdd indicates the week ending date.

The consolidated report is then distributed via email to every member of the team, as well as to the business, sales, and management teams and whoever else needs to know about project status.

Figure 10–5 Status report

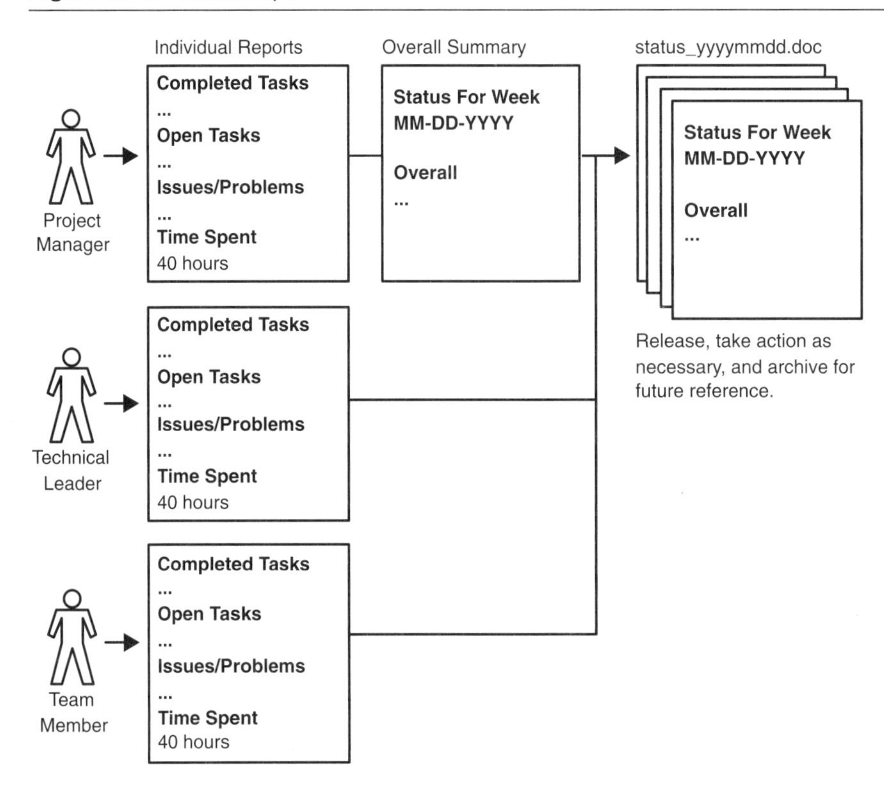

The project manager can also hold a weekly team status meeting to review the tasks completed, open tasks, and issues and problems. Action items (i.e., the open tasks) are acted upon by team members over the following week(s).

Finally, the project manager archives each status report for future reference in, for example, Microsoft Visual SourceSafe. As time progresses, the series of status reports will grow and can be referred to as needed.

10.1.5 Financials

There are many ways to determine the financial state of an application development project. One simple way is based on totaling up the hours each team member has spent on the project and using their hourly rate to calculate a dollar value.

The project manager is generally responsible for monitoring and managing the financials of the project. In complex, enterprise application development projects, there can be a large number of people and resources. However, there are also many tools available to help you manage the project financially, such as Microsoft Excel 2003.

Figure 10–6 Financial spreadsheet

	A	B	C	D
1	W/E	VL Hr	HS Hr	RS Hr
2	Aug/09/2003	40	40	40
3	Aug/16/2003	40	40	40
4	Aug/23/2003	40	40	40
5	Aug/30/2003	40	40	40
6	Sep/06/2003	40	40	40
7	Sep/13/2003	40	40	40
8	Sep/20/2003	40	40	40
9	Sep/27/2003	40	40	40
10	Oct/04/2003	40	40	40
11	Oct/11/2003	40	40	40
12	Oct/18/2003	40	40	40
13	Oct/25/2003	40	40	40
14	Nov/01/2003	40	40	40
15	Nov/08/2003	40	40	40
16	Total Used (Hr)	560	560	560
17	Total Remaining	0	0	0
18	Total $	84000	70000	70000
19				
20	VL $150 Per Hr * 560 Hrs			
21	HS $125 Per Hr * 560 Hrs			
22	RS $125 Per Hr * 560 Hrs			

For example, in Figure 10–6, we assume that there are three people on a project. Each team member has 560 hours allotted to them and is anticipated to work solidly for 40 hours a week over 14 weeks. The rate for the technical leader is $150 per hour, while the coders cost $125 per hour each. This gives you a total application development project value of $224,000.

This example is quite simple. In actuality, you may also need to calculate additional items such as material purchases, travel, vacation time, etc. Certain employees also may not be needed throughout an entire project. For example, you may not need to bring coders into the project until requirements gathering has been completed. Nonetheless, from spreadsheets such as Figure 10–6, you can calculate a variety of values (such as cost, margin, and profit) to help determine the project's progress and financial state.

10.1.6 Environment

In large enterprises, there may be several separate environments, such as Development, Integration, and Production environments (see Figure 10–7). The existence of separate environments allows you to develop and test new code prior to production deployment. This reduces the chances of errors appearing on production systems.

Ideally, the server hardware, software, and application configurations should be identical in all three environments. Practically, however, this is not usually possible. Full triplication is an expensive option, since you are tripling the number of servers and software.

Figure 10–7 Enterprise environments

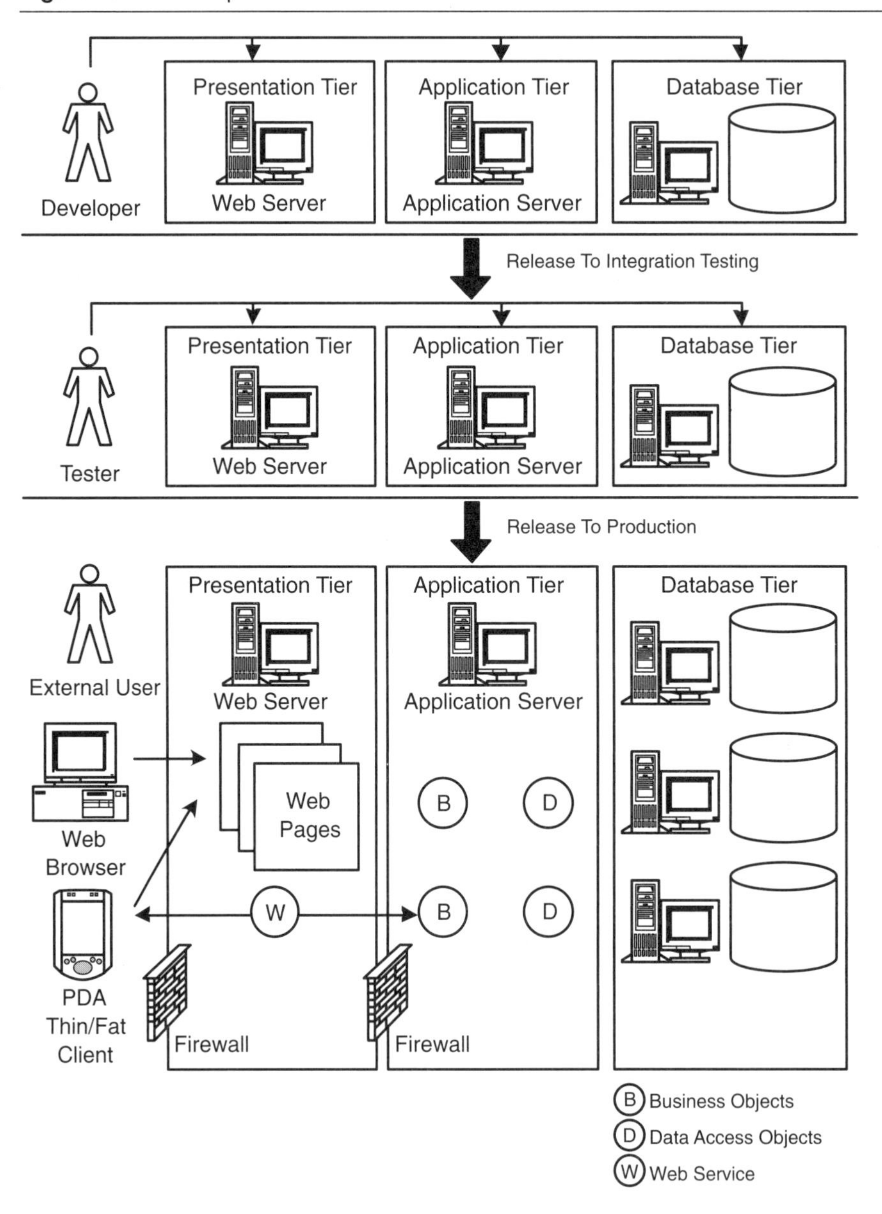

The following sections describe each environment in more detail.

Development. Code is initially created and modified using a set of tools on the development systems. Typically, the Development environment is functionally ahead of the Integration and Production Environments and is often a very dynamic environment.

The Development environment often does not have the same number of servers the Integration and Production environments have, and may consist of a variety of devices that even include developers' laptops and workstations.

Integration/quality assurance. After coding and unit testing is carried out in the Development environment, all necessary files are packaged into release form and placed onto the integration systems. Integration testing can then be carried out by an independent group. Once the code has passed integration testing, which includes functionality, performance and security tests, the code can be released to production.

The Integration environment is smaller than the Production environment in terms of numbers of servers but is functionally as similar as possible. It is typically much more stable and static than the Development environment. In very high-quality environments, there may be more than one Integration environment between the Development and Production environments.

Production. The production systems consist of a full complement of servers, software, and applications. Production systems are typically used and run by operations staff. The main operations in the production environment include handling periodic releases of packaged and application software, cleaning up files, re-indexing, running backups, and running the production systems. If you utilize good development practices, the Production environment should be very stable and operable.

It is important that no coding or development be done on either the integration or production systems. All development should be done in the Development environment, tested, and then released to the Integration and Production environments through a standard release.

Developers may or may not have access to the Integration or Production environments. While developers generally have access to all code and documents, they might not be given release privileges and must pass all developed code, documents, etc. to a separate group of testers for integration testing and/or to operators for production releases.

10.1.7 Version and Change Management

Project documentation and code needs to be managed and versioned, otherwise, it becomes nearly impossible to track changes.

Several industry standard tools are available to help you manage changes to documents and code. One that is used extensively is Microsoft Visual Source Safe (VSS), which allows users to store many different document and code types.

Source control software, such as VSS, allows the team to place all project documentation and code into a central repository. Each developer checks out the document or code module he or she wants to work on, makes the necessary changes or additions, and tests it. When the document is complete or the code is working, the developer checks the code back into the repository.

When it is time for a release, the latest version of the code is withdrawn from the repository, packaged, and released to the Integration or Production environments. After the release is done, the team labels the repository with the release number and the date. This version control system allows documents and code to be rolled back to previous versions if necessary.

10.1.8 Training

Large-scale projects may require that a variety of people be trained in the use of the new mobile application and technologies. These users include end users, integration testers, operators, help desk personnel, and administrators. There are many types of training available, from the very formal to the quite casual including:

- Formal classroom training
- Reading project documents (e.g., design, release, operations documents)
- Reading online documentation (e.g., help documentation)

10.1.9 Common Managerial Headaches

The following sections describe some of the more common managerial headaches.

Contract constraints. One interesting problem that can crop up is the scenario that occurs between pre-sales and implementation of a project. In order to win the opportunity to implement the project, you typically have to submit a proposal to a prospective customer. The proposal can be extremely detailed and consist of architectural details, as well as cost, resource, and timeline expectations. You might argue that you shouldn't have any architectural details at all in this proposal. But if you don't, how will you accurately estimate the time and cost of the project? This is a tricky scenario because you are bound by the proposal and contract metrics, which are typically negotiated before you have gathered your requirements and carried out the design. These requirements might end up being completely different than what was proposed.

Blending requirements and design. It is important to keep requirements and architectural design separate. It is quite common to inadvertently blend requirements with what is actually technical design. While it may be helpful to do this in order to explain functional requirements, it is important not to use requirements in lieu of design. For example, the following requirements have design considerations blended into them:

- "The architecture must support 100 Pocket PC users."
- "The system must support 802.11b."

In the first requirement, it is stated that the use of Pocket PC (Microsoft-based) products, is required. If your existing enterprise architecture is Java-based, this may not be the ideal solution. Your existing development and operations personnel may only be familiar with Java-based technologies, so you may actually want to use mobile devices that use J2ME instead.

In the second requirement, the specific protocol stated should ordinarily be identified at design time. For example, there actually may not be any requirement to follow 802.11b at all. Perhaps all that is actually required is that mobile users are continually connected to the server applications.

Choosing technologies too soon. It is extremely tempting to choose technologies and languages before requirements and design have been completed. For example, when you propose a project to a prospective customer, a list of hardware and software is often requested in order to help the customer better understand the cost of the project. Thus, you may specify the purchase of 100 HP iPAQ mobile devices. However, you are doing so before the requirements have actually been fully specified. Suppose, for example, that you subsequently find out that you actually need 50 HP iPAQs or indeed 50 Tablet PCs. You will have purchased too many devices and/or the wrong devices—a costly mistake.

As a rule of thumb, if you find you are discussing vendors, technologies, and languages at the requirements level, you may be getting too deep into the architectural design instead of formulating requirements.

10.2 REQUIREMENTS

Requirements are needed in order to help formulate and crystallize what the application must be able to do. There are many types of requirements and several ways to gather and document them. This section describes these items in more detail.

10.2.1 Requirement Types

The development of any mobile application must address several types of requirements. Below, we list some of the most common and important requirement types. This is only a synopsis of the full requirements that are necessary but serves to highlight the most important areas.

User requirements. User requirements typically specify a human's interaction with the system, the user interface, user documentation, and what the system look likes. Typically, all user types need to be considered, including end users, operators, help desk personnel, and administrators.

Business requirements. The proposed mobile application must be able to address business requirement questions including:

- **Business purpose.** It is important to bear in mind that a mobility solution is ultimately implemented to help solve a business problem. It must, therefore, have a sound business purpose.
- **Total Cost of Ownership (TCO).** If you intend to mobilize your existing applications and workforce, it is important to consider the TCO for a mobile device for each user. The TCO includes material costs (e.g., hardware, software, and network services), development costs (e.g., application management and development), and operation costs (e.g., operations, support, and training). Some of these are one-time costs (e.g., device purchases and development) but there are also recurring costs (e.g., wireless connectivity, support, and operational costs). The TCO can amount to as much as several thousand dollars per mobile device per year, depending on the configuration.
- **Total Benefit of Ownership (TBO).** Some things enterprises should look for to determine the TBO of a mobile application are increased worker productivity, increased revenue resulting from an increase in productivity, increased accuracy, increased safety, and decreased costs based on improving efficiency and eliminating redundant processes.
- **Return on Investment (ROI).** The ROI can be calculated once the TCO and TBO are well understood. Generally, the greater the capabilities, features, and applications the mobile device holds, the higher the TCO. However, these increased capabilities can also result in a large TBO, which can positively affect the ROI. As long as the ROI is positive, the proposed application is probably worth developing.

Functional requirements. The proposed application is typically required to perform a set of well-defined functions. These include:

- **Core functions.** Core functionality includes authentication, enrollment, and access control. Other functions, such as session management or state management, might also be required. Example functional requirements include "Users must be able to authenticate through the use of a login page" or "Users must be able to self-enroll to the application service."
- **Application-specific functions.** The application's specific functions depend on what the application actually does. In the case of banking or trading systems, for example, these functions include making trades, transferring funds, etc. Example application-specific requirements include "Users must be able to buy/sell stocks using a mobile device," "Users must be able to transfer funds from their checking account to their savings account over the Internet using a mobile device," and "The mobile device must support field personnel who do not have a network connection."

- **Utility functions.** Utility functions might include email or printing capabilities. Example utility functions include "Users must be able to print out reports" and "Users must be able to send secure email to administrators."

Operations requirements. Operations requirements pertain to the application environment, maintenance, and support. Typical operations requirements include:

- **Environment.** The requirements for the Development, Integration, and Production environments must be defined. For example, you may require that no source code be placed in the Production environment. You may also specify the database contents in all three environments be separate.
- **Maintenance and support.** These requirements specify how the application is to be maintained and supported after it has gone into production.

System requirements. A number of system requirements are typically considered, including:

- **Security.** The servers and applications may be required to meet certain security standards prescribed by the enterprise, the industry, or some other institution.
- **Performance.** The servers and applications may be required to meet certain performance requirements, such as availability, response time, and transaction rate.
- **Scalability.** The servers may be required to be both horizontally (more servers) and vertically (more powerful servers) scalable as usage, load, and demand increases.

Miscellaneous requirements. In addition to the basic set of requirements, application architectures are typically required to follow a set of miscellaneous requirements, such as:

- **Leading- or bleeding-edge technology?** In terms of software, remaining up to date and compliant with leading-edge technology and tools may be desirable. However, this option can be expensive, risky, and unnecessary, making it difficult to justify from a business perspective.
- **Personnel.** Companies do not necessarily have the personnel to handle extremely complex technology. As a result, standard, common, easily available, and inexpensive tools that are not too difficult to learn in a reasonable time frame are often used to create applications.
- **Standards compliance.** The design and deployment of the system may be required to adhere to certain recognized industry standards, such as Institute of Electrical and Electronic Engineers (IEEE), W3C, or government software development standards.
- **Openness and vendor neutrality.** Many companies are interested in open and vendor-neutral solutions, and may not consider proprietary solutions that tie them to a single vendor. Companies also are usually interested in solutions provided by vendors that are stable and have a reputation for excellence. Thus, an attractive solution might include HP or IBM hardware and Microsoft, Oracle, or Sun software.

10.2.2 Gathering Requirements

Gathering requirements for a mobile application development project can be done in a number of ways. The following sections describe some common methods.

1. **Ask key people.** One important way to gather requirements is simply to ask key people involved in the project, such as sales, marketing, business, and technical people. These key people are often referred to as *stakeholders* (since they have a vested interest in the success of the project) and include:

    - People who requested that the system be built (e.g., sales, business)
    - Future users of the system
    - People who work with the existing applications and infrastructure
    - Subject matter experts (e.g., mobility, web, legacy system experts)

    There are a variety of ways in which you ask key people questions about the requirements of the future application. Some are quite formal, such as a set interview. Others may be less formal, such as a casual discussion. The main concern is to ensure that all points of view are well represented.

2. **Peruse existing documents.** There may be existing documents that you can read to help formulate the new requirements. Documents that typically contain requirements include:

    - The Request for Proposal (RFP) document
    - The Proposal or Statement of Work (SOW) documents
    - Existing requirement documents
    - Business case studies

10.2.3 Documenting Requirements

Using the raw information collected during requirements gathering, you may develop a requirements specification document. The requirements specification is a high-level description of everything the system must do. Requirements should be stated simply, preferably with some tangible measure.

The requirements specification may include diagrams or UML models of the requirements. Use cases are also a useful tool when developing requirements. A use case consists of the following elements:

- Title
- Use case number/ID
- Brief description
- Actors—People and/or systems involved
- Pre-conditions—Items that must be true before entering the use case

Figure 10–8 Use case

Use Case	
Title	Authentication
ID	Use Case #1
Description	Login procedure
Actors	Customer, Authentication System
Pre-Conditions	Customer must have a username and password.
Task Sequence	Customer enters username. Customer enters password.
Exceptions	If username or password is wrong, customer is shown an error message and the login page is shown again. If username and password are correct, customer is allowed to enter main application.
Post-Conditions	Customer can see main application.

- Task sequence—List of steps performed by the actors
- Exceptions
- Post-conditions—State of the system and actors after completing the use case

For example, the use case in Figure 10–8 describes a situation in which a customer is attempting to log in to the main application via a login page. The use case describes the scenario and all the elements listed above.

Use cases can also be modeled using simple diagrams. These diagrams show the relationships among the actors, the system, and the various use cases.

There are many tools that can be used to document requirements, including Microsoft Word, Microsoft Visio, and IBM Rational Rose.

10.3 DESIGN

The role of design is to translate the requirements specifications for what the system should do into a specification for how the system will do it.

Mobile applications must be designed like any other complex enterprise application. You should develop an architectural design, a set of use cases, flow charts, object models, data models, and all of the other artifacts typically associated with technical design.

It is important to attain a state of balance in design and not overthink the problem. If you over design, you can end up with what is called "analysis paralysis" and stall the whole project. However, if you underdesign, the architecture may be flawed or insufficient to meet your business requirements.

10.3.1 Design Documentation

The design of the system is typically captured in a set of design specification documents that encompass high-level to detailed design information. The following sections describe one set of design documents that typically can be used for designing large-scale, enterprise applications.

Overall architecture document. You may create an overall design specification document that contains a high-level design of the entire system, including the new mobile application to be built and its interface to any existing applications. Within this design, you should include major architectural components, such as:

- User characteristics
- Client characteristics
- Server tiers (e.g., Presentation, Application, and Database)
- Software components (e.g., pages, objects)
- Hardware components (e.g., clients, servers, network)
- System interfaces
- Integration with existing back-end systems
- Security
- Products (e.g., Microsoft IIS, Apache, Oracle 9iAS)
- Technologies (e.g., .NET, COM, J2EE, EJB, CORBA)
- Programming languages (e.g., VB.NET, VC++, C#, Java)
- Number of clients, servers, and software in each tier
- Specification of the placement of servers between firewalls
- Specification of the failover procedure (e.g., database clusters)

Detailed design documents. You may also create a detailed design specification document for each major architectural component within the overall architecture. Each of these documents may include:

- User interface design (e.g., HTML, ASP.NET, JSP)
- Object models (high-level description of application objects)
- Class diagrams (details of class properties and methods)
- Database design (e.g., table and field definitions)

The design specification documents can be written in a variety of ways and typically consist of:

- Text
- Images
- Use cases
- Flow charts
- Object models

- Class diagrams
- Data models

There are many tools available to help you document the design, including Microsoft Word, Microsoft Visio, and IBM Rational Rose.

The structure of this book is actually an amalgamation of these documents and follows their general pattern. For example, we have discussed the overall architecture of mobile applications in one chapter. We have also discussed a breakdown of the major architectural components in several other chapters.

10.3.2 Design Considerations

The design considerations for mobile applications have already been discussed at length in this book. Some of the key considerations are reiterated below.

1. **Users.** All users and their roles and entitlements must be incorporated into the architectural design. The users are the final customers and have to be satisfied. A rapidly developed prototype (or initial release) that puts a working version of the mobile application into the user's hands early on can also be very helpful. The user interface must be intuitive, elegant, and easy to use. If you don't satisfy your users, they will not use your system!
2. **Mobile device types.** The end-user mobile devices have to be taken into consideration. For example, based on your requirements, will these devices be cellular telephones, PDAs, Tablet PCs, or Laptop PCs? Or will they be some mixture?
3. **Fat versus thin clients.** The design of the mobile client is highly dependent upon the type of connectivity. If the mobile users will have continuous connectivity, you may be able to develop a thin client. However, if the mobile users will be only partially connected, you may be required to develop a fat client.
4. **Authentication.** Authentication is almost certainly required on the mobile device itself, on the servers, or on both. The authentication mechanism that is designed must be robust and secure.
5. **Abstraction, layers, and tiers.** In order to build a scalable enterprise architecture, the use of abstraction, layers, and tiers is a critical part of the design process.
6. **Application-specific functionality.** Application-specific functionality must also be designed. For example, if you are attempting to design the scanning of inventory items from a mobile device to a back-end inventory management system, this has to be detailed in your design documentation.
7. **Existing systems integration.** It is important to consider the existing architecture before adding mobile components. Try to look at the technology neutrally, while considering what the enterprise already has, including commercial off-the-shelf products such as web or application server products.

10.4 CODE DEVELOPMENT AND INTEGRATION

Mobile application projects may require development of a new mobile client application. They may also require development and integration with existing server-side applications. For example, you may have to change an existing web application to support the use of mobile devices.

10.4.1 Coding

Application coding involves the creation and/or modification of several types of software. The following items may require coding.

1. **Mobile client application.** If the mobile client is a fat client, the application will typically be coded in a language such as Microsoft VB.NET or C#. The application will contain forms, business, data, and communication objects. If the mobile client is a thin client, the application web pages will typically be coded and hosted on the server.
2. **Presentation Tier code.** Web pages may or may not exist on the presentation server. If the mobile client is a fat client, any existing web pages typically will not require substantial changes. However, you may need to create one or several web services that manage synchronization of data from the mobile device. If the mobile client is a thin client, you may need to generate an additional set of web pages that can be viewed on the mobile client (e.g., HTML, ASP.NET, JSP pages). These are typically the small-screen version of the current web pages.
3. **Business Tier code.** Typically, if the architecture is well-abstracted, a set of business objects will exist on the application servers. These are written in any number of languages, including Microsoft VB.NET, C#, and Java. The objects typically will not require major changes, although there might be some minor changes made in order to interface to the web service on the presentation server.
4. **Database Tier and existing back-end system code.** The Database Tier and existing back-end system code typically will not require major changes, although there might be some minor changes necessary to accommodate the new web service.
5. **Test scripts.** Developers may be called upon to write scripts that help test the application. Ideally, however, it is better if the integration testing team actually writes the test scripts. However, this is not always possible and developers may have to undertake this task along with normal development duties.
6. **Release scripts.** Developers may also be called upon to write release scripts that help release the application from the Development environment to the Integration and Production environments. While large enterprises may have the integration testing team do this, often it is still up to the developers to ensure their code can be easily released to the Integration and Production environments.

A variety of powerful tools are available to help coders develop, unit test, and release code, including Microsoft Visual Studio .NET and Borland JBuilder.

Coders should follow good code development practices. For example, code should be checked into a code management system such as Microsoft Visual Source Safe. If code changes are needed, you should check out the appropriate code module, modify it, unit test it, and check it back into the code management system. You should also follow good practices for releasing code. If these procedures are not already established, we recommend that you establish them.

10.4.2 Unit Testing

Unit testing consists of testing the individual functional pieces of code. Many modern development tools (such as Microsoft Visual Studio .NET) allow you to step through the code, line by line, and examine variables and other items of interest within the code. Using these tools can help you find, diagnose, and fix bugs and problems.

10.5 INTEGRATION AND SYSTEM TESTING

A mobile application development project must allow sufficient time for complete integration and system testing of the new mobile client application, the infrastructure, and any changes to the server-side application. Integration testing typically involves testing substantial components of the application and architecture until they have been shown to work together as a whole.

It is useful to note that it usually takes less time to test resources under the enterprise's control. Testing resources not under the enterprise's control (e.g., a cellular network operated by a vendor) may take substantially longer to schedule and complete.

10.5.1 Testing Process

There are several approaches to integration testing including white or black box testing. Of the two methods, black box testing is probably the more common approach. White box testing tests a component's internals and is more detailed and time-consuming than black box testing, which allows you to just look at the input and output of a component without actually knowing precisely what is happening internally.

Essentially, black box testing supposes that you don't really care how it works as long as what you enter gives you a correct and predictable result. For example, if we test authentication using a black box approach, the input is the username and password (see Figure 10–9). The expected result or output is either access to an application or an error message and a request to re-enter the information. Internally, the authentication mechanism may be doing a substantial amount of work, authenticating against an access control database and a back-end system. However, because this is a "black box" to the tester, the tester may not need to know the details of the mechanism. All the tester needs to ensure is that the expected results are obtained.

Figure 10–9 Authentication black box testing

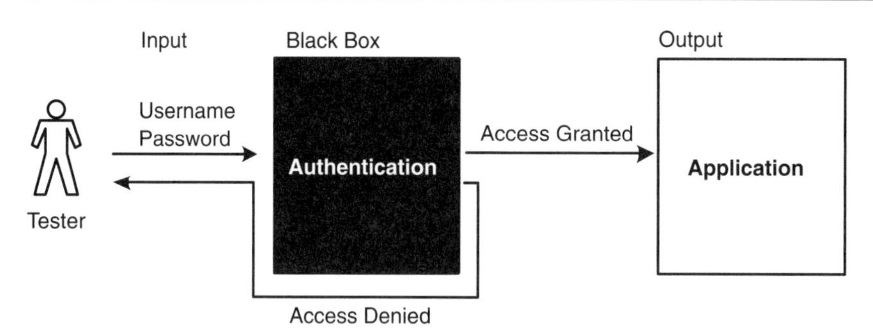

10.5.2 Testing Documentation

Tests are usually documented in a test specification consisting of multiple individual test cases (see Figure 10–10). Each test case tests a single test scenario; in large enterprises, there may be hundreds or thousands of test cases. The test case itself is typically composed of several items, as follows:

- Title
- Test case number/ID
- Brief description
- Actors—People and/or systems involved
- Pre-conditions—Conditions that are required to complete the test
- Task sequence—List of steps performed by the actors
- Exceptions
- Post-conditions—Conditions that exist after the test
- Data entered
- Actual result
- Passed/failed
- Tester name

The test cases may be partially or completely automated. Since integration testing is ongoing, it can be very efficient and helpful to code a set of test scripts that execute tests automatically. The test cases should be written by the integration testers. Developers may also be called upon to help write test scripts, but this is somewhat less ideal.

Figure 10–10 Test case

Test Case	
Title	Authentication
ID	Test Case #1
Description	Successful Login
Actors	Customer, Authentication System
Pre-Conditions	Customer must have a username and password.
Task Sequence	Customer enters username. Customer enters password.
Exceptions	If username or password is wrong, customer is shown an error message and the login page is shown again. If username and password are correct, customer is allowed to enter main application.
Post-Conditions	Customer can see main application.
Data Entered	
Actual Result	
Passed/Failed	
Tester Name	

10.5.3 Testing Considerations

Test cases should be carried out from several perspectives. These perspectives are described below:

Functionality and usability testing. Functionality testing is one of the main testing considerations, since it tests the system's functionality as a whole. Items to be tested include web pages, code executables, and documentation. In addition, the usability of the application—such as the user interface—is also tested.

Performance testing. Performance testing can be carried out at many levels. Typically, items that are tested include traffic load, number of users, application page access and utilization, and server uptime and downtime. Many tools are available for performance testing, including Microsoft Web Application Stress Tool and Mercury Interactive Load Runner.

Security testing. Security testing can also be carried out at many levels, including intrusion detection, firewall protection, port lockdown, anti-virus checking, encryption, and username and password protection. Again, there are many tools available, including the Check Point firewall, Microsoft ISA firewall, Norton AntiVirus, and McAfee Security Center.

Regression testing. Regression testing is carried out after new releases of code and documentation are finished. New functionality, performance, and security tests are run and compared against previously recorded test results to ensure that the results are both better than before and also do not break currently operational and functionally correct code.

10.6 DEPLOYMENT AND RELEASE MANAGEMENT

After having developed and tested pages and code on the development and integration testing systems, a well-documented process for assembling the release files and moving them to production should be implemented. Generally, a fallback procedure is also desirable, so that this process can be reversed if an emergency occurs and version reversion becomes necessary.

Deployment and release management may involve many clients and servers. One or a mixture of the following mechanisms may be used to move files from development systems through the integration testing systems and to the production systems:

- Automated release scripts and tools
- Zipped or compressed files
- CDs
- Shared disk drives for drag-and-drop copying

There are many file types that can be included in a release. Some of the most common include:

- Web page files (e.g., HTML, ASPX, JSP, XML)
- Executable code files (e.g., EXE, DLL)
- Text files (e.g., TXT)
- Image files (e.g., GIF, JPG, MPG)
- Documentation (e.g., Microsoft Word, Adobe PDF)

The following sections describe the deployment and release management of these files in more detail.

10.6.1 Mobile Client Production Releases

After the mobile application is developed and tested, it must be deployed to the Production environment. For fat client applications, this may involve deployment to many end-user mobile device types, including Pocket PCs, Tablet PCs, and Laptop PCs. If there are a large number of mobile devices to be set up and configured, a set of custom scripts may be written to help distribute the code. In contrast, thin clients are generally much simpler to deploy since there is no application software to be released to the client. This is one of the reasons thin clients are very appealing.

10.6.2 Server Production Releases

A release to the existing web application and back-end servers may also need to be performed. Typically, each of the tiers has slightly different release and setup procedures, as described below.

Presentation servers. Setting up the presentation servers typically involves installing and configuring a commercially available web server. Several are available for a variety of server platforms, including Microsoft IIS and Apache. The operators and administrators must have the knowledge and privileges associated with carrying out normal system administration in order to properly manage all web servers. If there are a large number of these servers, specially written scripts are typically used to set up the web servers and application software.

Application servers. Setting up the application servers typically involves installing and configuring an application server software product. Several are available for a variety of server platforms, including Microsoft IIS and Oracle 9iAS. If there are a large number of these servers, a set of custom scripts can be written to set up the custom application software.

Database servers. Setting up the database servers involves installing and configuring databases on a server. Some of the most popular databases include Microsoft SQL Server 2000 and Oracle 9i. Generally, there are not a large number of database servers and release scripts are not typically written. However, schema and table setup SQL scripts may be written to help accurately create, modify, and delete tables.

Existing back-end systems and databases. The existing back-end systems and databases may already be set up. If not, scripts may need to be written to enable their setup and configuration.

10.7 RE-EVALUATION AND REITERATION

If the iterative and incremental spirals model is used, after a release has been finished, there are typically two things you do. You evaluate (or re-evaluate) your situation and then reiterate. The following sections describe what is involved in these steps.

10.7.1 Re-Evaluation

The evaluation process after a release requires you to view what has been accomplished in the release and what is left to do. Several items can be produced, including:

- A list of functionality that has been deployed
- A list of requirements that the release's functionality has met
- A list of problems with the release

10.7.2 Reiteration

Reiteration simply involves returning to the planning and management phases. At this time, you may need to adjust your project plan according to what came to light during the evaluation process. You may also document problems in the status report. You should then continue going through the requirements, design, coding, and testing phases again until you reach your second release. You then re-evaluate and reiterate again.

With each progressive release, the total functionality left to deploy and the number of problems should decrease. If you find that each individual release progressively contains more functionality, this may be a sign of "scope creep" and you should attempt to trade off the resources, schedule, or features of the application. If there are progressively more functionality problems, you may need to verify and improve the design and code quality.

This process continues until you have a production-ready release. This occurs when all functionality has been deployed and is fully operational with no detectable errors. Alternatively, this occurs when you have no new functionality or fixes left to deploy.

10.8 OPERATIONS AND MAINTENANCE

At some point, a decision is made to actually "go into production." Even though you may have carried out releases to the Production environment systems for some time, you may not actually be running in production mode.

In order to do so, several items need to be considered. First, the code and documentation must be error-free and meet all requirements. Second, the integration testing team must be satisfied that the code and documentation is acceptable. Finally, the operations and administration staff must be able to operate, support, and maintain the applications and servers. This usually means that operators and administrators are well-trained and able to operate the application using a set of enterprise class tools and utilities. We will address some of the most important operations and maintenance considerations in the sections below.

10.8.1 Day-to-Day Operations

Operators and administrators need the ability to operate and work on the mobile client and server application. Often, this takes the form of an administration web application that is accessible using a mobile device or web browser. This web application may have two types of functionality, as described below.

1. **Base functionality.** This includes functions for enabling and disabling users, managing databases, handling production releases and performing backups. For example, an operator or administrator may need the ability to create a user profile.
2. **Impersonation functionality.** Operators and administrators may also need the ability to work on the mobile client and server applications as seen through the eyes of an end-user.

For example, if the new application pertains to entering a stock trade from a mobile device, an operator or administrator may need the ability to impersonate the user and enter a stock trade on the user's behalf.

10.8.2 Monitoring

Operators and administrators will normally monitor the applications and systems on an ongoing basis and act on certain information. Several typical types of monitoring are described below.

1. **System monitoring.** It is not very efficient or pleasant for human operators and administrators to have to continually monitor applications and systems. It is better to implement a monitoring tool that is capable of detecting a significant event (e.g., when the system is down), alerting an operator or administrator through an email or pager message, and taking some remedial action.

   For example, a separate monitoring system may periodically ping or send a message to the mobile and server applications to see if a response is elicited. If the monitoring system receives a response, it assumes the system is alive; otherwise it assumes the system is down. There are several tools that are able to perform this type of functionality, including Mercury Interactive SiteScope and HP OpenView.

2. **Performance monitoring.** Operators and administrators generally require the ability to monitor traffic and evaluate the mobile and server applications using information such as the number of hits, time spent on certain pages, heavy traffic times, frequency distributions, CPU utilization, and database utilization.

   There are a variety of enterprise-class performance monitoring tools, most of which are based on gathering raw information on a continuous basis, periodically analyzing the information, and then generating a performance report. Mercury Interactive SiteSeer and HP OpenView are two good examples of this.

3. **Security monitoring.** Operators and administrators need the ability to monitor traffic and view traffic that has been filtered out for security reasons in order to detect and prevent intrusion. A number of security monitoring tools are available, including firewall logs and virus detectors such as Norton AntiVirus or McAfee Security Center.

10.8.3 Backup and Restoration

Operators also require the ability to back up and restore the mobile and server application database and data. Backups are typically performed on an ongoing and regular basis (e.g., every day or every week). Backups are usually either full backups, which are complete copies of the database and file data, or incremental backups, which are a copy of the changes made since the last full backup. It is also important to periodically restore the backup data to another database to ensure that you can truly do so. You do not want to find out when your primary database is down that your backup copy also cannot be restored.

A wide variety of backup tools are available to help operators and administrators. Some are very simple while others are more sophisticated. For example, you can back up and restore a Microsoft SQL Server database and tables using the product's own backup and restore utility. In addition, there are standard backup utilities for Pocket PCs, Tablet PCs, and Laptop PCs. For more complex backup needs, there is software that allows you to back up and restore the entire contents of a Pocket PC, including registry settings, programs, files, and databases, to a Compact Flash card or some other backup medium.

10.9 SUMMARY

In this chapter, we applied a high-level approach to managing large-scale enterprise mobile application development projects. Project management of enterprise application development is a large topic unto itself and well beyond the scope of this book. Nonetheless, we felt it was important to discuss some of the aspects of large-scale enterprise application development management. We used an interpretation of the Rational Unified Process (RUP) and applied it to mobile application development management. Many large companies also provide formal project management methodologies and practices for planning, designing, developing, and deploying technology solutions (e.g., Microsoft Solutions Framework). Interested readers are referred to the references in the back of this book for further information.

CHAPTER 11

Mobile Museum Case Study

And therefore never send to know for whom the bell tolls;
It tolls for thee.
—John Donne

This case study discusses the mobilization of a hypothetical existing web site to accommodate mobile device users with a constant network connection. We start by describing the characteristics of the existing web site. We then describe the procedure we followed to mobilize the web site by developing smaller versions of the web pages to support Pocket PCs and other mobile devices. Readers who are interested in specific details of this case study may download the full code listing from the companion web site.

11.1 USE CASES

Suppose you are running the technology center of a large art museum and you already have an existing Internet web site that provides information about a variety of masterpiece paintings exhibited in the museum. The web site also allows users to register with the museum so that they may receive museum-related information or subscribe to various museum services.

You would like to provide museum visitors with a mobile device that contains the same information about the paintings as the original web site and also allows the gathering of the same visitor information. The mobile device will be wirelessly networked into the museum's network and a thin client will be used. As a result, mobile device users will not be able to see the museum's web pages if they are disconnected from the network.

A Pocket PC, such as an HP iPAQ running Microsoft Windows Mobile 2003 and Microsoft Pocket Internet Explorer, will be made available to visitors for a nominal charge. Visitors will rent the device at the museum kiosk, leave a credit card number as a deposit, and walk around the museum with the Pocket PC, which will give them information about particular masterpieces. The visitors can also fill out a Web Form to register for various museum services.

When a visitor is ready to leave, he/she will return the Pocket PC to the museum kiosk. Since the visitor was connected for the entire visit, any gathered information has already been entered into the server database. There is no need to synchronize data between the Pocket PC and the server.

As an alternative to renting a Pocket PC, a museum visitor can also use his/her own cellular telephone or other mobile device to browse the museum's web site as long as he/she has connectivity to the Internet.

11.1.1 Use Case Actors

The current museum web site supports users in two different roles (see Figure 11–1). When developing use cases for the mobile guide to the museum, we considered each of these roles and how they might be portrayed as actors in use cases. There are two types of actor as described below:

1. **Prospects.** Prospects are web site users who are able to view information about the paintings located in each gallery in the museum. They may also submit information about themselves and even make a donation to the museum. We refer to these users as *prospects* since each user is a prospective museum member or benefactor. One of the primary goals of this application is to allow prospects to enter their personal information, credit card information, and a donation amount.
2. **Administrators.** Administrators are able to add, edit, and delete the data submitted by prospects, as well as update and maintain gallery and painting information that is stored in the web site's database.

The following sections describe use cases involving each of the two actors.

Figure 11–1 Use case actors

11.1.2 Existing Web Site Use Cases

The existing web site supports a number of use cases. Some of these may be performed by prospects who visit the web site from their home computers; others may be performed by administrators who work at the museum. Prospects may perform the following tasks on the web site (see Figure 11–2):

View galleries. All web site users may view a list of galleries in the museum and select a specific gallery to view its details. Gallery details include a map that shows the number and location of paintings within the gallery.

View paintings. All users may view the paintings that exist within a specific gallery. Each painting's web page includes digital photographs of the painting along with the painting's title, description, and artist information.

View help. All users may view a simple but comprehensive help page, which provides helpful information about the museum and use of the web site.

Submit personal information. All users may submit their personal information in order to subscribe to various museum services. If users want to make a donation, they may also submit their credit card information and a gift amount.

Figure 11–2 Prospect use cases

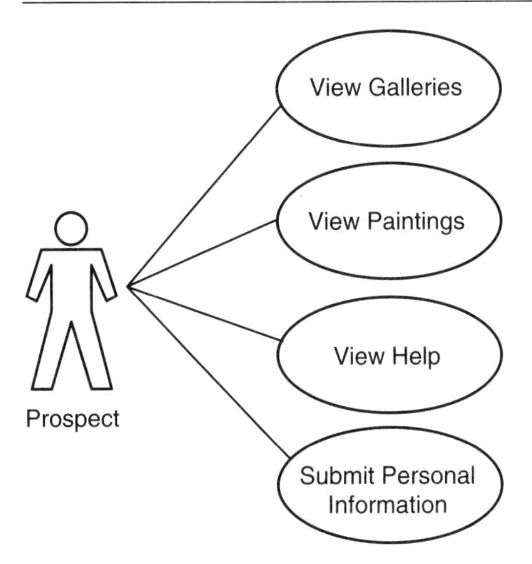

Administrators may perform a variety of additional tasks on the existing web site (see Figure 11–3).

Log in. Administrators must enter an ID and password to log in to the administrative area of the web site.

Add/edit/delete users. Administrators may add, edit, and delete web site user accounts, which are stored in the web site's database.

Add/edit/delete roles. Administrators may add, edit, and delete web site roles, which are stored in the web site's database.

Add/edit/delete galleries. Administrators may add, edit, and delete gallery information, which is stored in the web site's database.

Add/edit/delete paintings. Administrator may add, edit, and delete painting information, which is stored in the web site's database.

Add/edit/delete prospect information. Administrators may add, edit, and delete the information entered by prospective museum members. For example, administrators may

Figure 11–3 Administrator use cases

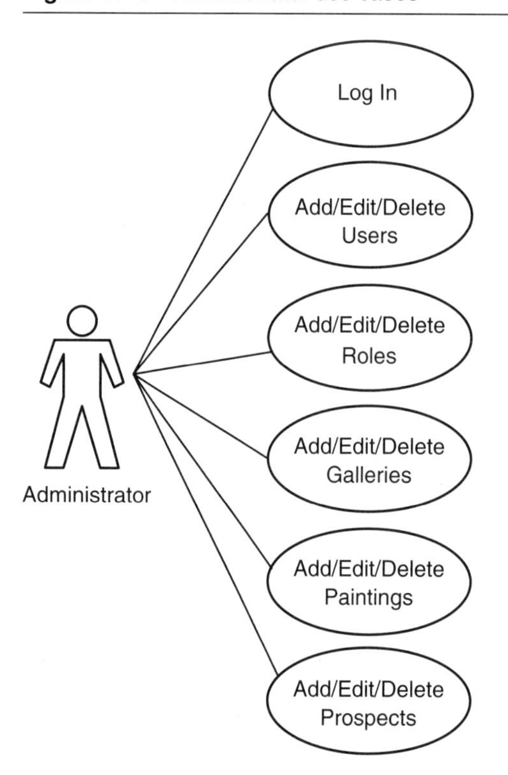

visit the web site each day to see how many new prospective members have expressed an interest in the museum or have made a donation.

11.1.3 Mobile Web Site Use Cases

When you mobilize an existing application, you must decide which use cases to support in the mobile client application. In this case, we decided to support all the prospect use cases. However, we decided not to mobilize any of the administrative use cases, since these functions will typically be performed by an administrator who is working in an office using a desktop or laptop computer. The supported use cases are listed again below:

View galleries. We thought it was important to support this use case since a primary goal of the mobile application is to present a "Mobile Museum Guide," which would be useless without gallery information.

View paintings. This use case is also crucial to the development of the Mobile Museum Guide.

View help. Although not strictly necessary, we decided to keep this use case since it is very simple and should be easy to support in a mobile application. In addition, this may allow us to provide help specifically tailored to the mobile web site.

Submit personal information. We decided to keep this use case because one of the main goals of creating the Mobile Museum Guide is to provide a useful service to visitors that will hopefully entice them to make a donation to the museum. Visitors are more likely to donate if they can make a donation on their mobile devices before leaving the museum rather than waiting until they get home.

11.2 ARCHITECTURE

The architecture of the Mobile Museum Guide application is illustrated in Figure 11–4. The mobile client consists of the following items:

- Pocket Internet Explorer Web Browser (Pocket PC)
- WAP Browser (cellular telephone)

The existing web application consists of the following items:

- ASP.NET Web Forms
- Business Objects
- Data Access Objects
- Database

Figure 11–4 Mobile Museum Guide architecture

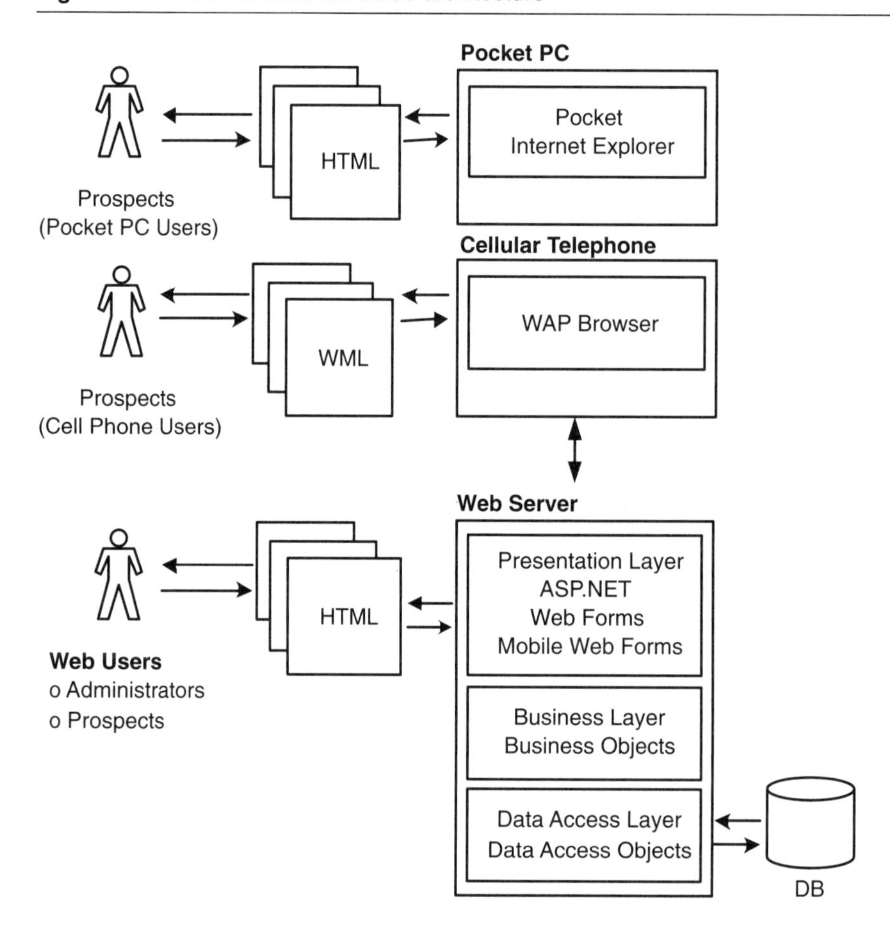

In addition, the mobilized web application contains the following items developed to support mobile users:

- ASP.NET Web Forms (to support Pocket PCs)
- ASP.NET Mobile Web Forms (to support cellular telephones)

These items are discussed in detail throughout the rest of this chapter.

11.3 CLIENT DETAILED DESIGN

As this is a thin client application, there is no custom application code on the client. The client is also assumed to always be connected. As a result, we do not need to consider storing any data or files on the mobile device, and a simple web or WAP browser will suffice for application display.

In our case study, a HP iPAQ Pocket PC running Pocket Internet Explorer would be rented out to museum visitors. However, we did not want to restrict visitors and prospective members from using their own devices so we developed the application to support any mobile device with a web or WAP browser. We have tested this sample application using the following mobile devices:

1 Pocket PC running Microsoft Windows Mobile 2003
2. Pocket PC Emulator running Pocket PC 2003
3. Microsoft Mobile Explorer 3.0
4. Nokia Mobile Browser Simulator 4.0

This is not intended to be a comprehensive list of devices supported by this application; other devices make work just as well.

11.4 SERVER DETAILED DESIGN

This section describes the detailed design of the current web application's Presentation, Business, and Data Access Layers using class diagrams and object models. All of the objects on the server are in the namespace "Mobile Museum." We have created several different sub-namespaces to hold the various objects. Figure 11–5 illustrates the namespaces in the existing Mobile Museum web server application.

11.4.1 Existing Presentation Layer

In our case study, the existing presentation layer consists of a set of ASP.NET web pages and user controls, as well as a set of JPG images. These are described below.

1. **Utility web pages and controls.** Figure 11–6 illustrates some of the utility web pages and controls, which are described below:

   Global. This page replaces the traditional global.asa file in ASP applications. The Global class provides event handlers such as Session_Start, Application_Start, Application_AuthenticateRequest, etc. The Application_AuthenticateRequest method creates a new Custom Principal object and adds it to the current HTTP Context.

Figure 11–5 Namespaces in the Mobile Museum server application

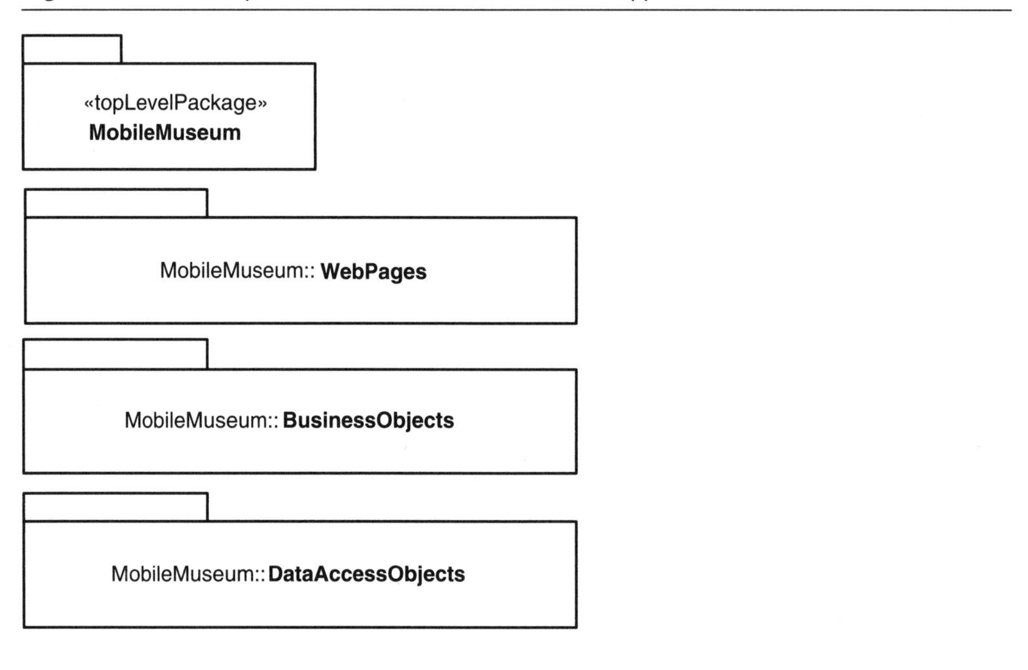

Figure 11–6 Utility web pages and controls

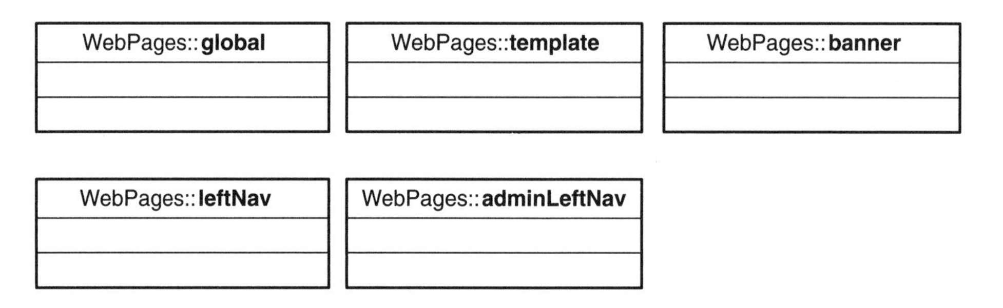

Template. This is an ASP.NET page that we used to develop the user interface for the web site. Every time we created a new ASP.NET page, we used this page as a template, cutting and pasting its HTML into the new page.

Banner. This web user control displays a banner at the top of the page.

LeftNav. This web user control displays the left-hand navigation bar for the unsecured area of the web site.

AdminLeftNav. This web user control displays the left-hand navigation bar for the secure administrative area of the web site.

2. **Default Web pages.** Figure 11–7 illustrates the default web pages that are available to all users. The existing web site is protected by Microsoft "Forms Authentication." However, the default pages can be accessed by anyone. This is specified in the Web.config file as follows:

```
<authentication mode="Forms">
<forms loginUrl="/MM_WebSite/login.aspx"
name="sqlAuthCookie" timeout="60" path="/"></forms>
</authentication>

<authorization>
<allow users="*" /> <!— Allow all users —>
</authorization>
```

The default web pages are described below:

Default page. This is the home page of the site.

Help. This page displays help for using the web site.

Signup. This page allows visitors to sign up for museum services.

Vw_galleries. This page shows a list of galleries in the museum.

Vw_gallery. This page shows a list of paintings in a particular gallery.

Vw_painting. This page displays a painting's details, including digital photographs of the painting and the painting's title, description, and artist information.

Figure 11–7 Default web pages

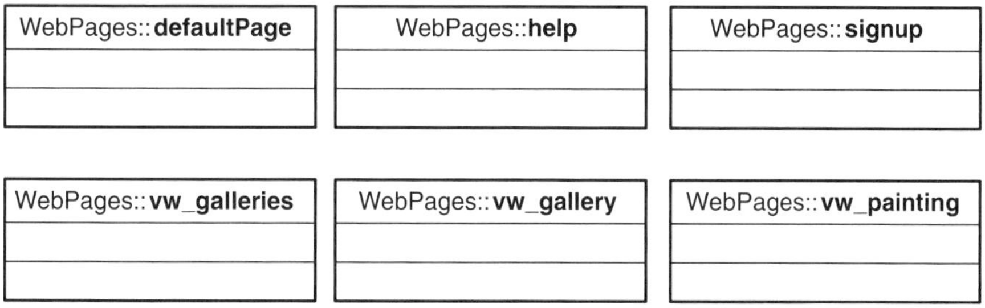

WebPages::**defaultPage**

WebPages::**help**

WebPages::**signup**

WebPages::**vw_galleries**

WebPages::**vw_gallery**

WebPages::**vw_painting**

3. **Administrative web pages.** There are several pages that are only accessible to administrators (see Figure 11–8). These pages are used to add, update, and delete users, roles, galleries, and paintings. To secure these pages, we placed them in a sub-directory called "admin." We also placed a Web.config file in that directory, which indicates that the pages are only accessible to users in the role "ADMIN."

```
<authorization>
<deny users="?"/>
<allow roles="ADMIN"/>
<deny roles="DEFAULT"/>
</authorization>
```

The web pages that are only accessible to administrators are as follows:

Login. Although this page is accessible to all users, it is only used by administrators, since they are the only ones required to log in.

The login page allows administrators to enter a username and password. If the information entered is valid, the page issues the Forms Authentication token. It also sets a cookie containing user information. This cookie is used by the Global class's Application_AuthenticateRequest method to create the Custom Principal object. The Custom Principal object is used by the Forms Authentication filter to perform role-based security.

The login page is also used to log administrators out of the web site by calling the FormsAuthentication.SignOut method, deleting the user information cookie, and removing the user from the HTTP Context.

Admin home. This is the home page of the administration section of the site.

Users. This page allows the administrator to view a list of users.

Roles. This page allows the administrator to view a list of roles.

Prospects. This page allows the administrator to view a list of prospects who have submitted information through the web site.

Edit user. This page allows the administrator to add or edit a user's information, including the user's role.

Edit role. This page allows the administrator to add or edit role information, such as its name and description.

Edit prospect. This page allows the administrator to add or edit a prospect's information.

View gallery. This page shows the details of a gallery, including its list of paintings.

Edit gallery. This page allows the administrator to edit basic gallery information.

Figure 11–8 Administrator web pages

WebPages::**login**	WebPages::**admin_home**	WebPages::**users**
WebPages::**roles**	WebPages::**prospects**	WebPages::**edit_user**
WebPages::**edit_role**	WebPages::**edit_prospect**	WebPages::**view_gallery**
WebPages::**edit_gallery**	WebPages::**edit_galleries**	WebPages::**edit_painting**

Edit galleries. This page allows the administrator to edit the list of galleries.

Edit painting. This page allows the administrator to add or edit a painting's information.

11.4.2 Business Logic Layer

This section describes the objects in the MobileMuseum.BusinessObjects namespace.

1. **MMUser class.** This class represents a user of the Mobile Museum application (see Figure 11–9). It has the following public properties:

   Id. This is an integer that uniquely identifies the user.

   FirstName. This is the user's first name.

   LastName. This is the user's last name.

   Name. This is a combination of the user's first and last name. It is read-only and is used for display purposes.

   RoleId. This is the identifier of the user's role.

   RoleName. This is the name of the user's role.

   Email. This is the user's email address.

   Username. This is the username used to log in to the web site.

Password. This is the user's password. For simplicity, we did not encrypt the password for this sample application. However, in a production application, we suggest encrypting the user's password using the Forms Authentication object's HashPasswordForStoringInConfigFile method.

The MMUser class has the following public methods:

New. This method is the default constructor.

New (Id). This constructor queries the database to populate all of the fields based on the user's identifier, which is passed as a parameter.

Delete. This method deletes the user from the database.

GetUserList. This method returns an array list of MMUser objects representing every user in the database.

Login. This method verifies the username and password against the value in the database.

NameIsTaken. This method is used to check if a particular username has already been taken. It is useful for preventing duplicates whenever you are adding a new user or changing an existing user's username.

Save. This method saves the user's data to the database. If the user Id is zero, the method inserts the record, retrieves the new user Id and populates it. If the user Id is greater than zero, the method performs an update.

Figure 11–9 MMUser class diagram

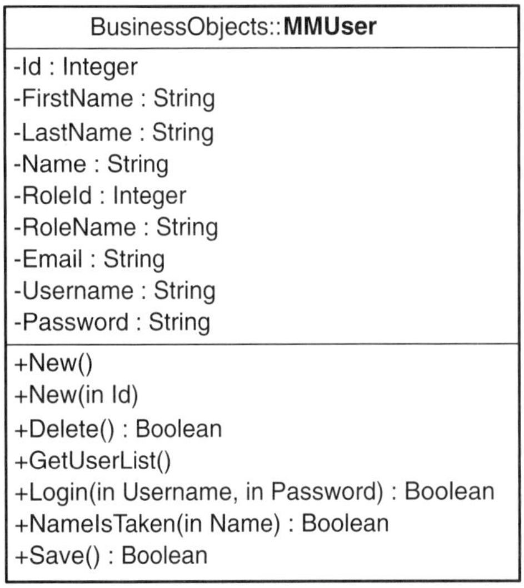

BusinessObjects::**MMUser**
-Id : Integer -FirstName : String -LastName : String -Name : String -RoleId : Integer -RoleName : String -Email : String -Username : String -Password : String
+New() +New(in Id) +Delete() : Boolean +GetUserList() +Login(in Username, in Password) : Boolean +NameIsTaken(in Name) : Boolean +Save() : Boolean

2. **MMRole class.** The MMRole class represents a security role for the Mobile Museum web application (see Figure 11–10). It has the following public properties:

   Id. This is an integer that uniquely identifies the role.

   Name. This is the name of the role, such as "Default" or "Administrator."

   Description. This is a brief description of the role.

   The MMRole class has the following public methods:

   New. This method is the default constructor.

   New (Id). This constructor queries the database to populate all of the fields based on the role's identifier, which is passed as a parameter.

   Delete. This method deletes the role from the database.

   GetRoleIdByName. This method looks up the identifier for a role given its name.

   GetRoleList. This method returns an array list of MMRole objects, representing every role in the database.

   NameIsTaken. This method is used to check if a particular role name has already been taken. It is useful for preventing duplicates whenever you are adding a new role or changing an existing role's name.

   Save. This method saves the role to the database. If the role Id is zero, the method inserts the record, retrieves the new role Id, and populates it. If the role Id is greater than zero, the method performs an update.

Figure 11–10 MMRole class diagram

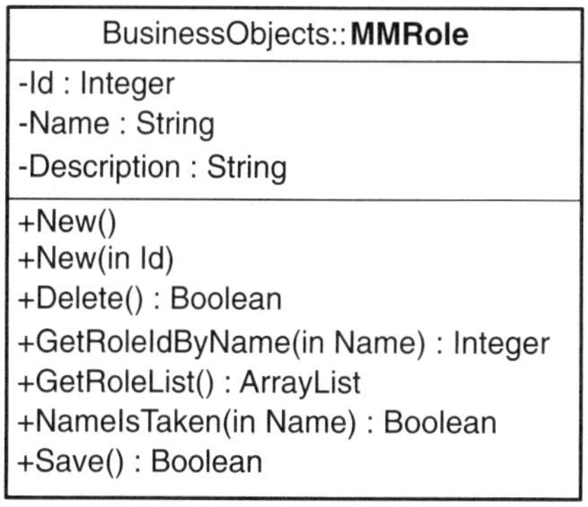

BusinessObjects::**MMRole**
-Id : Integer -Name : String -Description : String
+New() +New(in Id) +Delete() : Boolean +GetRoleIdByName(in Name) : Integer +GetRoleList() : ArrayList +NameIsTaken(in Name) : Boolean +Save() : Boolean

3. **Custom Principal class.** The Custom Principal class is used for web site security (see Figure 11–11). It implements the IPrincipal interface and is used for .NET roles authorization. Although the Microsoft .NET Framework comes with a Generic Principal object, we decided to create our own Principal object and implementation of the various properties and methods.

   When a user logs in, he/she is issued a Forms Authentication token and a cookie with certain information, such as his/her user identifier, username, name (first name and last name), and role. This cookie is used in the Global class's Application_AuthenticateRequest event. If the user is authenticated, the information is extracted from the cookie and is used to create a new Custom Principal object. This object is then placed into the HTTP Context for that web session. Our Custom Principal class contains the following public properties:

   Id. This is the identifier of the user who logged in.

   Identity. This object represents the identity of the user who logged in, including the user's identifier, name, username, and role name. This is read-only.

   Name. This is the first name and last name of the logged-in user.

   RoleName. This is the name of the logged-in user's role.

   Username. This is the username of the logged-in user.

   The Custom Principal class contains the following public methods:

   New. This method is the default constructor.

   New (Identity, Id, RoleName, Name, Username). This constructor populates all of the Custom Principal class's public properties.

   IsInRole. This method checks to see if the logged-in user is in the requested role.

Figure 11–11 Custom Principal class diagram

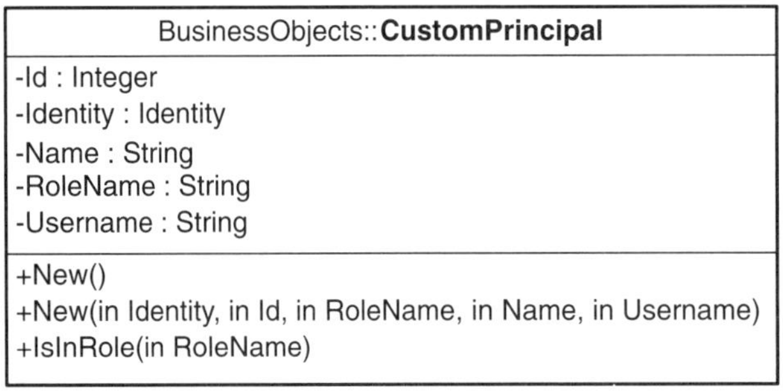

BusinessObjects::**CustomPrincipal**
-Id : Integer -Identity : Identity -Name : String -RoleName : String -Username : String
+New() +New(in Identity, in Id, in RoleName, in Name, in Username) +IsInRole(in RoleName)

4. **Prospect class.** The Prospect class represents a visitor to the museum web site who is a prospective member or benefactor (see Figure 11–12). It contains the following public properties:

    Id. This is an integer that uniquely identifies the prospect.

    FirstName. This is the prospect's first name.

    LastName. This is the prospect's last name.

    MiddleInitial. This is the prospect's middle initial.

    Address. This is the prospect's address.

    Apartment. This is the prospect's apartment number.

    City. This is the prospect's city.

    State. This is the prospect's state.

    Zip. This is the prospect's ZIP code.

    Country. This is the prospect's country of residence.

    DaytimePhone. This is the prospect's phone number.

    EmailAddress. This is the prospect's email address.

    GiftAmount. This is the amount of the prospect's donation to the museum.

    DataComplete. This is a flag indicating whether or not the prospect's data is complete.

    AccountNumber. This is the prospect's credit card account number.

    CreditCard. This is the prospect's credit card type, such as Visa or MasterCard.

    ExpDateMonth. This is the month the prospect's credit card expires.

    ExpDateYear. This is the year the prospect's credit card expires.

    LastUpdateDate. This is the date on which the prospect's record was last updated.

    The Prospect class contains the following public methods:

    New. This method is the default constructor.

    New (Id). This constructor queries the local database to populate all of the fields based on the prospect's identifier, which is passed as a parameter.

    Delete. This method deletes the prospect from the database.

    GetProspectList. This method returns an array list containing all of the prospects found in the database.

    Save. This method saves the prospect to the database. It performs an insert if the Id is zero, or an update if the Id is greater than zero.

Figure 11–12 Prospect class diagram

BusinessObjects::**Prospect**
-Id : Integer -FirstName : String -LastName : String -MiddleInitial : String -Address : String -Apartment : String -City : String -State : String -Zip : String -Country : String -DaytimePhone : String -EmailAddress : String -GiftAmount : String -DataComplete : Boolean -AccountNumber : String -CreditCard : String -ExpDateMonth : Integer -ExpDateYear : Integer -LastUpdateDate : String
+New() +New(in Id) +Delete() : Boolean +GetProspectList() : ArrayList +Save() : Boolean

5. **Gallery class.** The Gallery class represents a gallery in the museum, which contains paintings (see Figure 11–13). It contains the following public properties:

   Id. This is an integer that uniquely identifies the gallery.

   Number. This is the number used to refer to the gallery, such as 1, 2, 3, etc.

   Name. This is the name of the gallery, such as "20th Century Art," "Ancient Egyptian Art," etc.

   MainImage. This is the name of the main image file shown on the web page for this gallery. This is usually a map of the gallery.

   SmallImage. This is the name of the small image file shown in the side panel on the web page for this gallery.

   Paintings. This is an array list of Painting objects representing the paintings that hang in this gallery.

Figure 11–13 Gallery class diagram

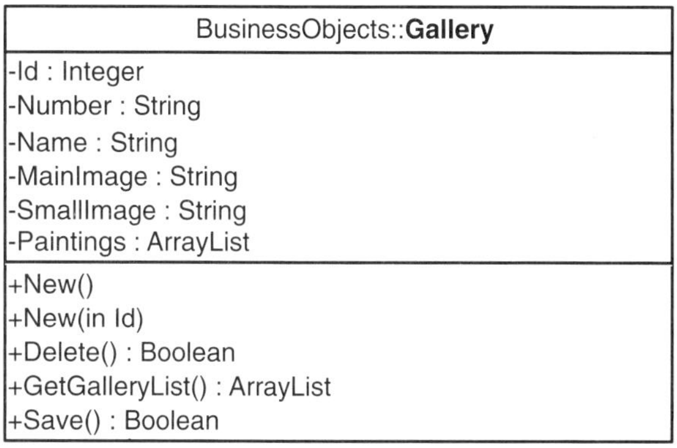

BusinessObjects::**Gallery**
-Id : Integer -Number : String -Name : String -MainImage : String -SmallImage : String -Paintings : ArrayList
+New() +New(in Id) +Delete() : Boolean +GetGalleryList() : ArrayList +Save() : Boolean

The Gallery class contains the following public methods:

New. This method is the default constructor.

New (Id). This constructor queries the local database to populate all of the fields based on the gallery identifier, which is passed as a parameter.

Delete. This method permanently deletes the gallery from the database.

GetGalleryList. This method returns an array list containing all of the galleries found in the database.

Save. This method saves the gallery information to the database. It performs an insert if the Id is zero, or an update if the Id is greater than zero.

6. **Painting class.** The Painting class represents a painting that hangs in a gallery in the museum (see Figure 11–14). It contains the following public properties:

Id. This is an integer that uniquely identifies the painting.

GalleryId. This is the identifier of the gallery to which this painting belongs.

Number. This is the number used to refer to the painting within the gallery, such as 1, 2, 3, etc.

Title. This is the title of the painting.

Painter. This is the name of the painting's artist.

Dates. These are the dates during which the artist lived.

Description. This is a brief description of the painting.

MainImage. This is the name of the main image file shown on the web page for this painting.

SmallImage. This is the name of the small image file shown in the side panel on the web page for this painting.

Figure 11–14 Painting class diagram

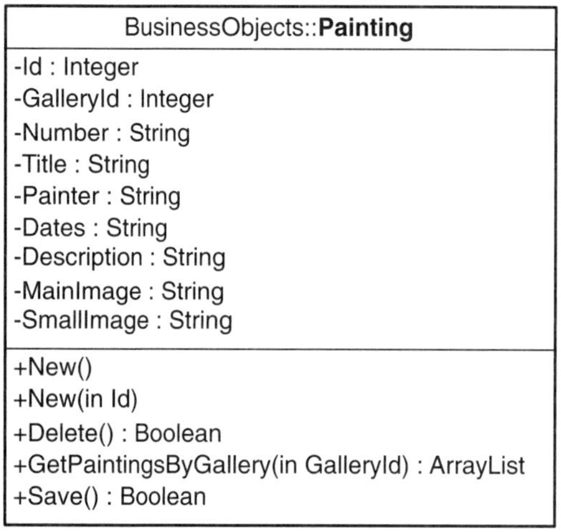

BusinessObjects::**Painting**
-Id : Integer -GalleryId : Integer -Number : String -Title : String -Painter : String -Dates : String -Description : String -MainImage : String -SmallImage : String
+New() +New(in Id) +Delete() : Boolean +GetPaintingsByGallery(in GalleryId) : ArrayList +Save() : Boolean

The Painting class contains the following public methods:

New. This method is the default constructor.

New (Id). This constructor queries the local database to populate all of the fields based on the painting's identifier, which is passed as a parameter.

Delete. This method permanently deletes the painting from the database.

GetPaintingsByGallery (GalleryId). This method returns an array list of all paintings located in the gallery with the specified identifier.

Save. This method saves the painting information to the database. It performs an insert if the Id is zero, or an update if the Id is greater than zero.

11.4.3 Data Access Layer

This section describes the database access objects.

1. **Data access objects.** All database access is performed by classes in the MobileMuseum.DataAccessObjects namespace (see Figure 11–15). The DataConnHelper class is used to look up the database connection string, which is then passed by the business objects to the SQLHelper class. The SQLHelper and SQLParameterCache classes are part of the Data Access Application Block for .NET and are provided by Microsoft for general use. Microsoft has developed several useful reusable application blocks to handle common tasks such as error handling and data access. For further information, see the Microsoft Patterns and Practices web site for a list of all application building blocks.

Figure 11–15 Mobile Museum data access objects

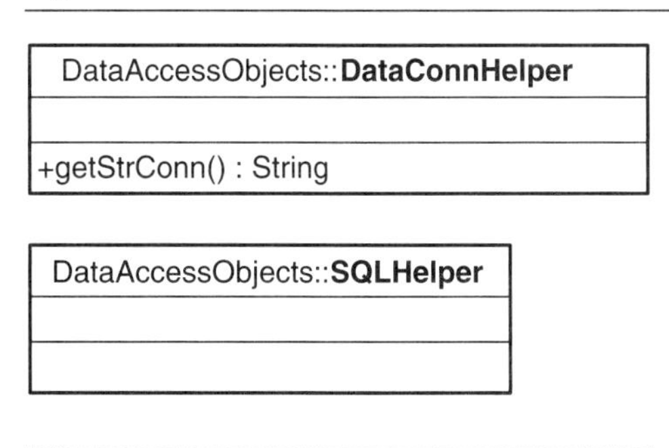

2. **Database design.** The server database was implemented using Microsoft SQL Server 2000. Figure 11–16 illustrates the database tables.

Figure 11–16 Database tables

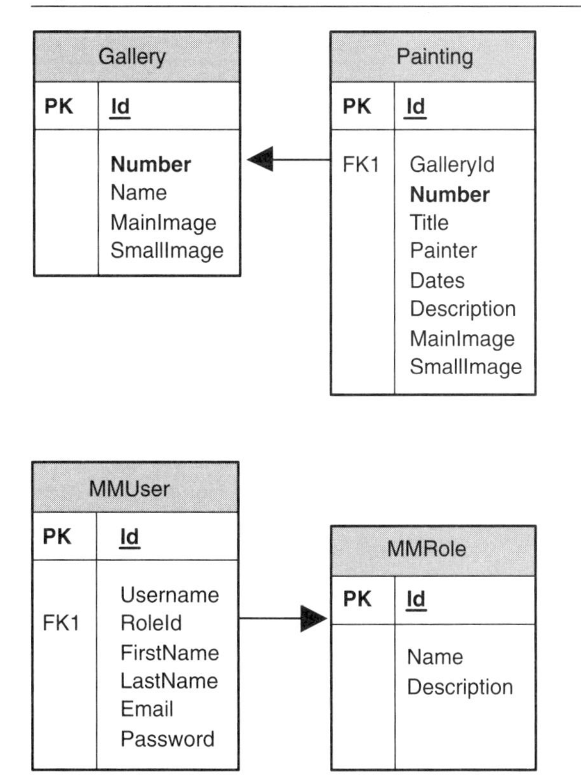

The Gallery table contains the following fields:

Field Name	Data Type	Primary Key
Id	int	Yes
Number	varchar(10)	
Name	varchar(255)	
MainImage	varchar(255)	
SmallImage	varchar(255)	

The Painting table contains the following fields:

Field Name	Data Type	Primary Key
Id	int	Yes
GalleryId	int	Foreign Key
Number	varchar(10)	
Title	varchar(255)	
Painter	varchar(255)	
Dates	varchar(255)	
Description	varchar(255)	
MainImage	varchar(255)	
SmallImage	varchar(255)	

The Prospect table contains the following fields:

Field Name	Data Type	Primary Key
Id	int	Yes
FirstName	varchar(50)	Foreign Key
LastName	varchar(50)	
MiddleInitial	varchar(50)	
Address	varchar(50)	
Apartment	varchar(50)	
City	varchar(50)	
State	varchar(50)	
Zip	varchar(50)	
Country	varchar(50)	
DaytimePhone	varchar(50)	
EmailAddress	varchar(255)	
GiftAmount	varchar(50)	
DataComplete	bit	
AccountNumber	varchar(50)	
CreditCard	varchar(50)	
ExpDateMonth	int	
ExpDateYear	int	
LastUpdateDate	smalldatetime	

The MMUser table contains the following fields:

Field Name	Data Type	Primary Key
Id	int	Yes
Username	varchar(50)	
RoleId	int	Foreign Key
FirstName	varchar(50)	
LastName	varchar(50)	
Email	varchar(255)	
Password	varchar(50)	

The MMRole table contains the following fields:

Field Name	Data Type	Primary Key
Id	int	Yes
Name	varchar(50)	
Description	varchar(255)	

The database contains the following stored procedures:

Procedure Name	Parameters	Return Value
sp_AddMMRole	@Name varchar(50), @Description varchar(255)	Id
sp_AddMMUser	@Username varchar(50), @FirstName varchar(50), @LastName varchar(50), @Email varchar(255), @Password varchar(50), @RoleId int	Id
sp_AddGallery	@Number varchar(10), @Name varchar(255), @MainImage varchar(255), @SmallImage varchar(255)	Id
sp_AddPainting	@GalleryId int, @Number varchar(10), @Title varchar(255), @Painter varchar(255), @Dates varchar(255), @Description varchar(255), @MainImage varchar(255), @SmallImage varchar(255)	Id

Procedure Name	Parameters	Return Value
sp_AddProspect	@FirstName varchar(50), @LastName varchar(50), @MiddleInitial varchar(50), @Address varchar(50), @Apartment varchar(50), @City varchar(50), @State varchar(50), @Zip varchar(50), @Country varchar(50), @DaytimePhone varchar(50), @EmailAddress varchar(255), @GiftAmount varchar(50), @DataComplete bit, @AccountNumber varchar(50), @CreditCard varchar(50), @ExpDateMonth int, @ExpDateYear int, @LastUpdateDate smalldatetime	Id
sp_UpdateMMRole	@Id int, @Name varchar(50), @Description varchar(255)	
sp_UpdateMMUser	@Id int, @Username varchar(50), @FirstName varchar(50), @LastName varchar(50), @Email varchar(255), @Password varchar(50), @RoleId int	
sp_UpdateGallery	@Id int, @Number varchar(10), @Name varchar(255), @MainImage varchar(255), @SmallImage varchar(255)	
sp_UpdatePainting	@Id int, @GalleryId int, @Number varchar(10), @Title varchar(255), @Painter varchar(255), @Dates varchar(255), @Description varchar(255), @MainImage varchar(255), @SmallImage varchar(255)	

Procedure Name	Parameters	Return Value
sp_UpdateProspect	@Id int, @FirstName varchar(50), @LastName varchar(50), @MiddleInitial varchar(50), @Address varchar(50), @Apartment varchar(50), @City varchar(50), @State varchar(50), @Zip varchar(50), @Country varchar(50), @DaytimePhone varchar(50), @EmailAddress varchar(255), @GiftAmount varchar(50), @DataComplete bit, @AccountNumber varchar(50), @CreditCard varchar(50), @ExpDateMonth int, @ExpDateYear int, @LastUpdateDate smalldatetime	

11.5 MOBILIZING THE EXISTING APPLICATION

This section describes the various considerations we made when mobilizing the existing application. It also discusses the detailed design of the new components we added to support mobile devices.

11.5.1 Architecture

As previously discussed, our existing application had a three-layer architecture, consisting of:

- ASP.NET Web Forms
- VB.NET Business Objects
- SQL Server 2000 Database

Because our existing application had a three-layered architecture, we only needed to make changes to the Presentation Layer. We were able to reuse the existing business objects and database without making any changes. We added a second web site to support Pocket PCs and other mobile devices. This new site contains a new set of ASP.NET Web Forms and Mobile Web Forms.

11.5.2 Users and Roles

Again, the existing application supports Prospects and Administrator roles. As we discussed earlier, there are different use cases associated with each of these roles. We decided to support all of the prospect use cases on the mobile device because we wanted prospects to have full functionality when using the Pocket PC rented at the museum. We especially wanted prospects to be able to view all of the paintings and to sign up and make a donation if they wanted to.

We decided not to support the administrator use cases on the mobile device since most of these administrative tasks are more easily completed by an administrator sitting at his/her desk and not roaming around the museum.

11.5.3 Presentation Layer

Even state-of-the-art mobile devices are still quite limited in terms of presenting web site page content. As a result, many existing web sites that are both functional and handsome on desktop or laptop devices are not rendered well on mobile devices.

In our case, these existing web applications needed the most work because the museum's existing web site was not very usable from a Pocket PC perspective and did not look very good in an unmodified state.

Figures 11–17 and 11–18 show the existing web site viewed on a laptop computer using Microsoft Internet Explorer. Figures 11–19 and 11–20 show the same pages viewed on a Pocket PC. As you can clearly see, the pages are not very attractive or easy to use, since they require both horizontal and vertical scrolling. This is certainly not likely to encourage visitors to browse the web site and donate money to the museum.

In order to alleviate this problem, we created a second web site to support the mobile devices. The detailed design of that web site is discussed below. We also made one change to the default page of the existing web site. We decided to automatically redirect users to the new site if they visit the normal web site with a mobile device. The page load event of the default page now contains the following code:

```
If Request.Browser("IsMobileDevice") Then
 Response.Redirect("/MM_MobileWebSite/default.aspx")
Else
 'check to see if it's Pocket PC 2003
 'by checking for IE and WinCE
 Dim capabilities As System.Web.Mobile.MobileCapabilities
 capabilities = CType(Request.Browser,
        System.Web.Mobile.MobileCapabilities)
 If capabilities.HasCapability("Browser", "IE") _
 And capabilities.HasCapability("Platform", "WinCE") Then
        Response.Redirect("/MM_MobileWebSite/ppc/default_ppc.aspx")
 End If
End If
```

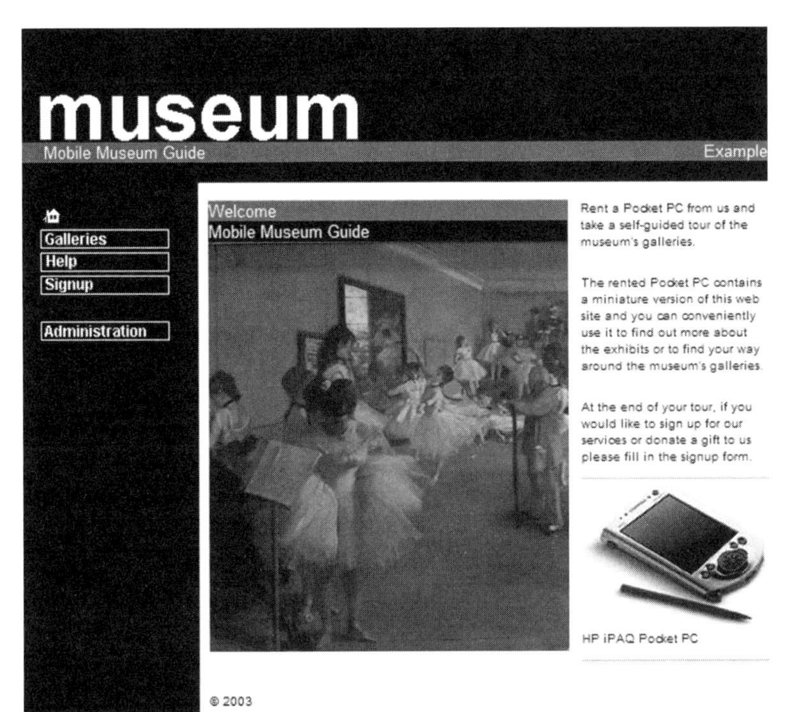

Figure 11–17
Museum web site home page

Figure 11–18
Museum web site signup page

Figure 11–19 Home page viewed with Pocket PC

The advantage of this redirect is that we do not need to publicize a new URL for our mobile web site. Users may continue to visit the museum's web site via the usual URL and they will be redirected to the new site only if necessary.

We also needed to change the JPG images on the web site, since they were too large to display on a Pocket PC screen. In addition, some mobile devices do not support JPG images. As a result, we made a second set of JPG images with the same names but with smaller dimensions. We also saved a copy of each small image as a GIF to support other mobile devices.

Figure 11–20 Signup page viewed with Pocket PC

11.5.4 Business Logic Layer

Our existing application contained the following business objects:

- MMUser
- MMRole
- Custom Principal

- Prospect
- Gallery
- Painting

We did not need to make any changes to these objects to support the mobile devices.

11.5.5 Data Access Layer

The existing database already stored user, role, and application data. We did not need to make any changes to support the mobile device.

11.5.6 Authentication

The existing web site only required authentication for administrators who wanted to access the secure pages. Since we decided not to support those administrative use cases on the mobile device, we did not need to support any authentication mechanism on our mobile web site.

However, if we had decided to allow administrators to perform secure tasks on a mobile device, we would have used Microsoft Forms Authentication on the mobile web site. The process would then be very similar to using Forms Authentication on the existing web site.

11.5.7 Enrollment

The existing web site provided an enrollment mechanism for prospects to sign up with the museum and make a donation. We decided to support this functionality on the mobile devices. This gives museum visitors the opportunity to sign up while they are still at the museum rather than waiting until they get home.

11.5.8 Administration

As previously discussed, the existing application provided several administrative functions that we decided not to support on the mobile web site.

11.5.9 Mobile Device Support

In this case study, the museum is specifically renting Pocket PCs to people and so it was of primary importance that the Pocket PC version of the web site be as attractive as possible. We also decided to support cellular telephones with WAP browsers to allow more mobile device types to be used. As a result, we developed an HTML template that we wanted to use for the Pocket PC version of the web site (see Figure 11–21).

Unfortunately, we could not create the same look and feel using the mobile controls that come with Microsoft Visual Studio.NET, which are necessary to support non-Pocket PC devices. To achieve this design, we needed HTML-specific formatting codes, such as tables, which are not

Figure 11–21 Pocket PC web site

permitted in Mobile Web Forms. We also wanted to use JPG images, which are not supported on certain mobile devices, such as cellular telephones.

Figure 11–22 shows our initial attempt at developing a sophisticated look and feel using the mobile controls. As you can see, the look and feel is not as good and the images are not as clear as in Figure 11–21 since we had to convert the JPG images to GIF images.

Figure 11–22 Web site using mobile Web Form controls

To alleviate this problem, we made two separate sets of pages; one that is used for the Pocket PC and the other that is used for cellular telephones. While this may seem like a lot of extra work, we felt it was necessary to achieve our desired look and feel on both device types. Thus, the Pocket PC web site uses standard HTML and ASP.NET while the cellular telephone web site uses the ASP.NET mobile controls. Figures 11-23 and 11-24 show the cellular telephone version of the web site viewed from the Nokia Mobile Browser Simulator 4.0.

When users visit or are redirected to the mobile web site, their browser capabilities are checked. If they are using Pocket Internet Explorer, an additional redirect is made to the Pocket PC-supported pages:

```
Dim capabilities As System.Web.Mobile.MobileCapabilities
capabilities = CType(Request.Browser,
  System.Web.Mobile.MobileCapabilities)
If capabilities.HasCapability("Browser", "Pocket IE") _
Or capabilities.HasCapability("Browser", "IE") Then
 Response.Redirect("/MM_MobileWebSite/ppc/default_ppc.aspx")
End If
```

Figure 11–23 Mobile web site viewed with phone simulator

Figure 11–24 Signup page viewed with phone simulator

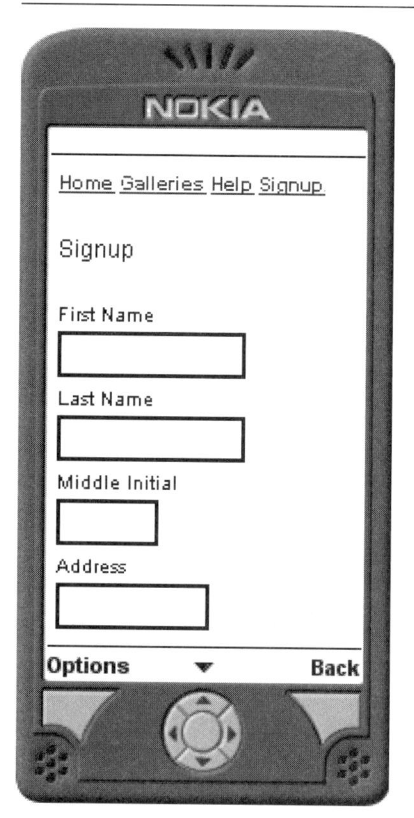

11.5.10 Presentation Layer Detailed Design

This section describes the detailed design of the new Presentation Layer components that we added to support mobile devices.

Figure 11–25 shows the namespaces that now exist in the Mobile Museum application including the new namespace MobileMuseum::WebPages::Mobile.

In our case study, the new Presentation Layer consists of a set of ASP.NET Web Forms and Mobile Web Forms (see Figure 11–26) as described below.

1. **Mobile Web Forms.** The new mobile web site contains only one mobile page called Default. This page contains several different mobile forms. As the user navigates through the site, different forms are activated and displayed including:

    frmWelcome. This form is the home page of the mobile site.

    frmHelp. This form displays help information.

    frmGalleries. This form displays a list of galleries from the database.

Figure 11–25 Namespaces in the Mobile Museum server application

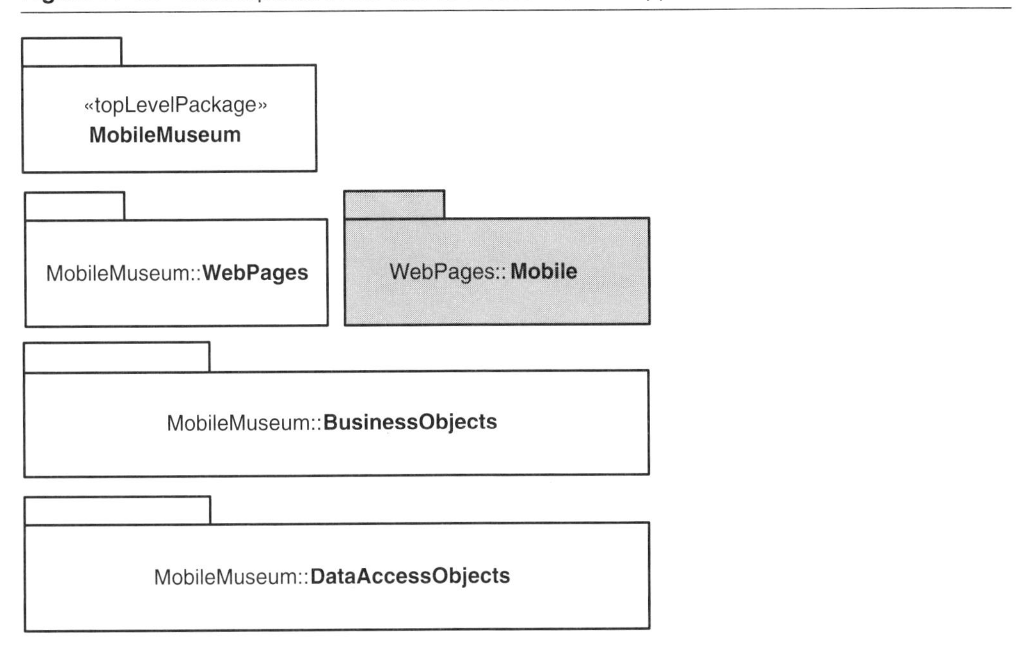

Figure 11–26 Mobile web site pages

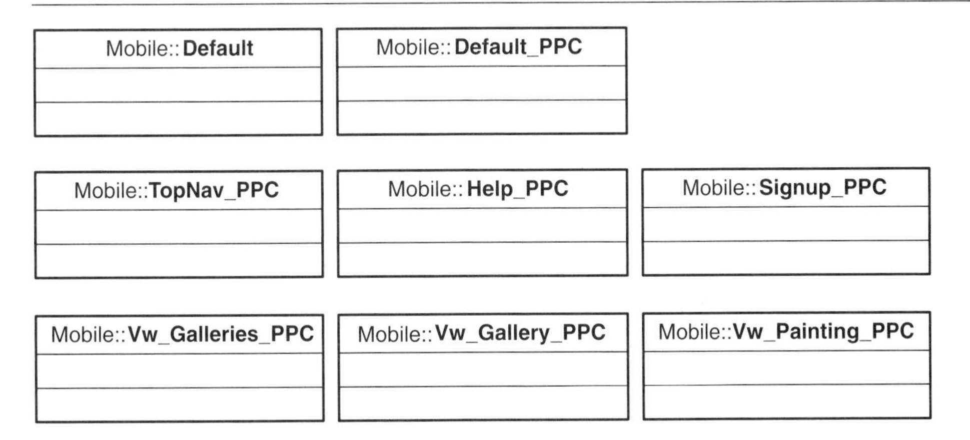

frmPainting. When a user first navigates to this form, it shows the details for the particular gallery, including a list of paintings in that gallery. Once the user picks a painting, it shows the details for that painting.

frmSignup. This form allows the user to sign up as a museum member and/or make a donation.

2. **Pocket PC web pages.** The new mobile web site contains several pages that are used only by the Pocket PC. These are normal ASP.NET Web Forms, whose HTML fits within the standard screen of a Pocket PC. These are not Mobile Web Forms and do not use the mobile controls; they are only intended for use by a Pocket PC. These pages include:

   Default_PPC. This is the default page of the Pocket PC version of the web site.

   Help_PPC. This is the help page of the Pocket PC web site.

   Signup_PPC. This is the signup page of the Pocket PC web site.

   TopNav_PPC. This is a web user control that displays the top navigation on each page of the Pocket PC web site.

   Vw_Galleries_PPC. This page displays a list of galleries.

   Vw_Gallery_PPC. This page shows the details of a gallery, including a list of paintings.

   Vw_Painting_PPC. This page shows the details of a particular painting.

11.6 DISCUSSION

The Mobile Museum Guide application provides an example of mobilizing an existing web application so that it supports Pocket PCs, cellular telephones, and other mobile devices. In this case study, users were expected to have a constant network connection, and so we did not need to develop a fat client application for the mobile device. We merely had to create new web sites on the server that support a variety of mobile devices.

This scenario can be applied to a variety of situations in which a company has an existing web site that needs to be mobilized and is able to provide a constant network connection.

11.7 EXTENSIONS

This web application could be further extended by adding location-sensitive devices that can detect a user's location and invoke an appropriate web page or respond accordingly.

One benefit of making this improvement is to make the web site more intelligent, so that it does not necessarily require human intervention to take action. For example, you can pinpoint the precise location of an indoor user using Newbury Networks' "Locale" products, and automatically invoke web pages depending on the user's location. Thus, in our example, it would be possible to automatically invoke the correct page when the user is walking past a particular painting without the user having to click on a link. This is a very handy feature (especially in environments where typing is more awkward or dangerous than in a museum).

11.8 SUMMARY

In our case study, we created two new sets of web pages. One set is used with the Pocket PC, which contains a relatively sophisticated HTML web browser. The other set of web pages supports cellular telephones that use WAP browsers.

The effort involved in reworking existing web sites to allow mobile users to conveniently utilize them may be substantial. The effort level will probably be greatest in one- or two-layer web applications where presentation and business logic is intertwined, and in web applications that require sophisticated HTML pages.

In our case study, we were fortunate because our existing architecture was three-layered. We were able to reuse our existing business objects and database and only had to rework the Presentation Layer.

CHAPTER 12

Mobile Biologist Case Study

Come wander with me, she said, into regions yet untrod;
And read what is still unread, in the manuscripts of God.

—Henry Wadsworth Longfellow

In situations where connectivity between a mobile client and a server cannot be guaranteed, it is often helpful to implement a fat client application. A fat client application is typically a simplified version of the full application on the server. It is used to gather and store data when the mobile client is disconnected from the server and synchronize that data with the server when connectivity is re-established.

For this case study, we have developed a sample application that runs on Pocket PCs and Tablet PCs. The application is called the "Mobile Biologist" and can be used by field biologists who regularly take water samples at a variety of sites and conduct various tests on water quality.

The Mobile Biologist application allows field biologists to visit water sources, sample the water, and record the results of the various water quality tests on their mobile device while they are still in the field. Since connectivity cannot be guaranteed while the biologists are out in the field, the application must be capable of storing information temporarily until connectivity with the server can be re-established.

For our purposes, a "water source" is a body of water that is regularly visited and tested by field biologists. For example, a water source may be the Central Park Reservoir or the East River in New York City. Each water source is identified by a name and a location, as well as latitude and longitude coordinates.

A "sample" is a water sample taken from a water source at a given time. Each sample is a record of the date, water temperature, and pH level of the water source. It is also a record of the levels of ammonia, nitrite, and nitrate in the water.

Each day, a list of water sources to be sampled is initially downloaded to the mobile device for fieldwork during the day. A biologist then visits each of the water sources, takes a water sample, and measures the various water parameters using a water test kit. The biologist then writes up the sample test measurements by inputting them into the mobile device. The measurements are stored locally on the device. Upon returning to the office, the biologist can synchronize his/her mobile device with the server database by uploading the results and thus updating the server database's water source history.

The Mobile Biologist mobile application is supported by a web site that allows biologists to add, edit, and delete water source and sample data. The web site also allows administrators to add, edit, and delete users. All site users may view reports on the water quality data.

Readers who are interested in the specific details of this case study may download the full code listing from the companion web site.

12.1 USE CASES

When developing this case study, we began by analyzing the requirements. Since this is a hypothetical example, we developed the requirements ourselves based on some investigations into aquatic biology and water testing. We have documented many of the requirements for the Mobile Biologist application as UML use case diagrams.

Under normal circumstances, we would typically gather requirements by interviewing the end users. For example, if a government agency was responsible for carrying out the water sampling, we might interview the agency's field biologists to ascertain their requirements.

12.1.1 Use Case Actors

The Mobile Biologist application supports users in multiple roles. When developing use cases, we considered each of these roles and how they might be portrayed as actors in use cases (see Figure 12–1). Here, there are three types of actor, as described below:

1. **Field Biologists.** Field biologists are able to add, edit, and delete water sources and samples, and view reports. For the purposes of this case study, only field biologists may use the mobile application. Field biologists have a choice of using either a Pocket PC or a Tablet PC, depending on personal preference.

Figure 12–1 Use case actors

Administrator

Default User

2. **Administrators.** Administrators are able to add, edit, and delete other users. They can also add, edit, and delete water sources and samples, and view reports.
3. **Default users.** Default users are only able to view reports.

The following sections describe the use cases involving each of the various actors.

12.1.2 Mobile Client Use Cases

The following use cases all involve the field biologist using the mobile device while connected to the network (see Figure 12–2):

Log in. The biologist may log in to the remote server database by entering a username and password. A copy of the username and password is saved in the local database so that the user can log in locally later. The biologist must successfully log in remotely at least once before he/she can work offline.

Download water sources and samples. The user may download all the existing water sources and samples from the remote database.

Upload changes. The user may upload all new sample data to the remote database.

The biologist may perform a limited number of actions on the mobile device while not connected to the network (see Figure 12–3):

Log in locally. The biologist may log in to the local database using the username and password that were saved when he/she first logged in to the server. Again, this cannot be done unless the biologist has successfully logged in remotely at least once.

Add samples locally. The biologist may add new water sample data to the local database.

Figure 12–2 Field biologist using mobile device connected to network

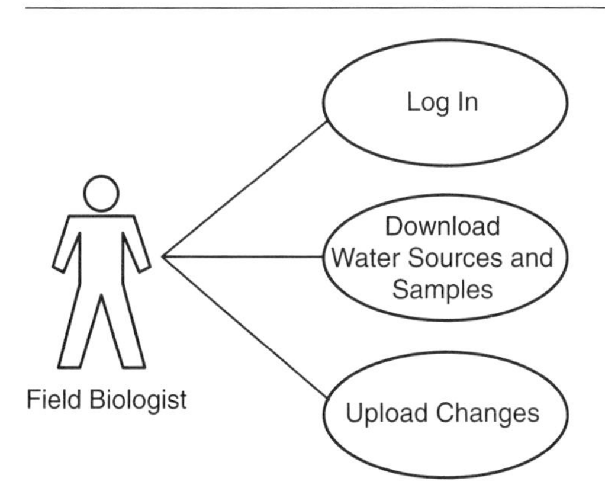

Figure 12–3 Field biologist using mobile device not connected to network

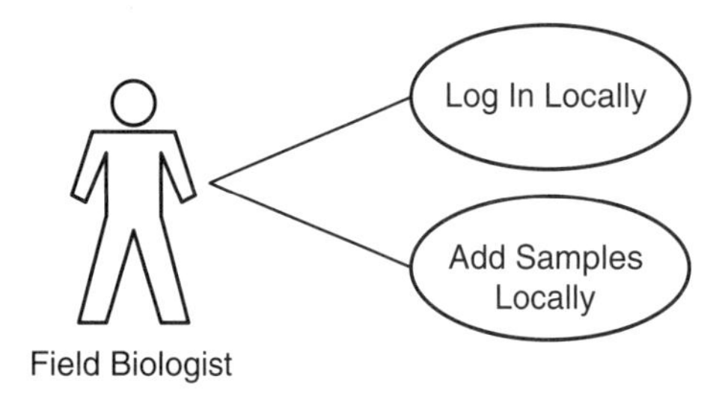

12.1.3 Web Site Use Cases

The following use cases involve the use of the Mobile Biologist web site and require a constant network connection. These use cases will typically be performed by one of the actors while working in an office and using a desktop PC to visit the web site (see Figure 12–4).

For example, a field biologist might visit the web site in the morning to update information on the server before downloading data to his/her mobile device and going into the field. The field biologist may perform the following actions on the web site:

Log in. The biologist may log in to the web site by entering a username and password.

Add/edit/delete water sources. The biologist may add, edit, and delete water sources using the web site. However, we do not allow the biologist to perform these actions on the

Figure 12–4 Field biologist using web site

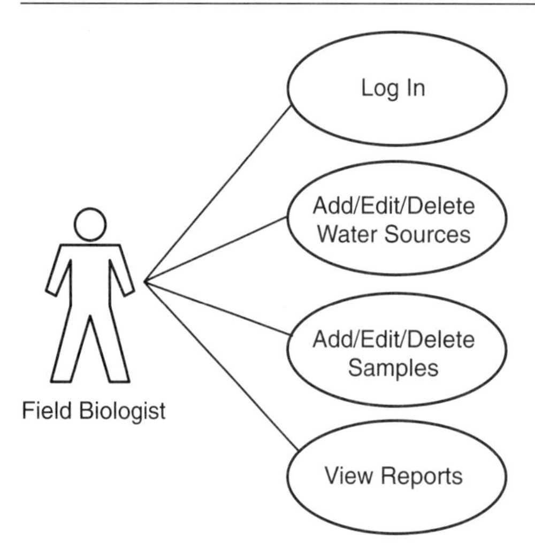

mobile device. Therefore, any water sources that need to be sampled must first be added to the web site before downloading data to the mobile device and going into the field.

Add/edit/delete samples. The biologist may add, edit, and delete samples. For this case study, we only allow the biologists to add samples on the mobile device, but he/she may add, edit, and delete samples using the web site.

View reports. The biologist can view water quality reports.

An administrator can perform the same functions as a biologist, as well as several additional functions (see Figure 12–5):

Add/edit/delete users. The administrator can add, edit, and delete users.

Add/edit/delete roles. The administrator can add, edit, and delete roles. He/she can also assign users to a role.

Default users are those users who have not been assigned any specialized role (such as Biologist or Administrator). Default users of the web site have very limited functionality and can only log in and view reports (see Figure 12–6).

Figure 12–5 Administrator using web site

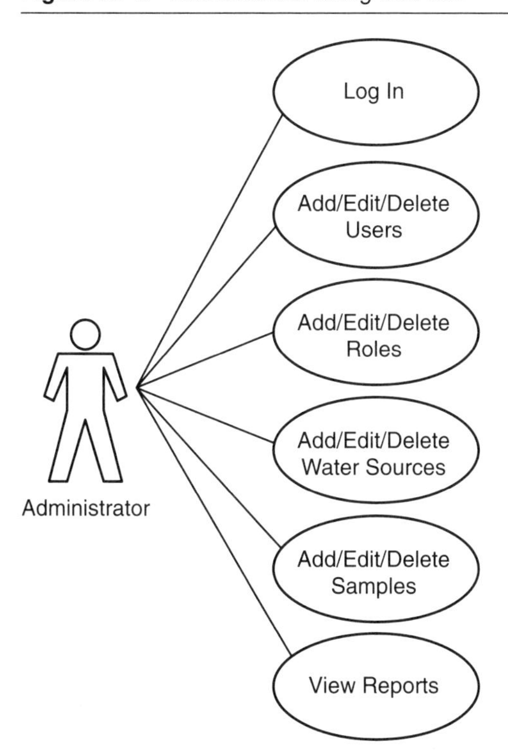

Figure 12–6 Default user using web site

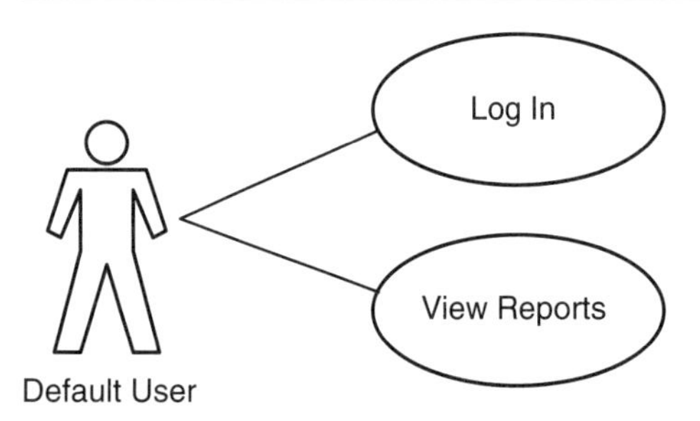

12.2 ARCHITECTURE

In the following sections we describe the architecture and flow of the Mobile Biologist application. Please note that we have implemented a simple web site for demonstration purposes only. In enterprise-wide web sites, multiple servers, security, multi-tiered web architectures, and other considerations must be taken into account.

12.2.1 Overall Architecture

The architecture of the application is illustrated in Figure 12–7. It illustrates that field biologists can use either a Pocket PC or a Tablet PC. The main difference between the two is that the Pocket PC runs the Microsoft .NET Compact Framework while the Tablet PC runs the full .NET Framework. Each application contains the following items:

- VB.NET Windows Forms
- Business Objects
- Data Access Objects
- Database
- Either .NET Compact Framework or .NET Framework

The Web Server contains the following items:

- HTML Content
- ASP.NET Web Forms
- XML Web Service
- Business Objects
- Data Access Objects
- Database
- .NET Framework

These components are all discussed in detail throughout the rest of this chapter.

Figure 12–7 Mobile Biologist architecture

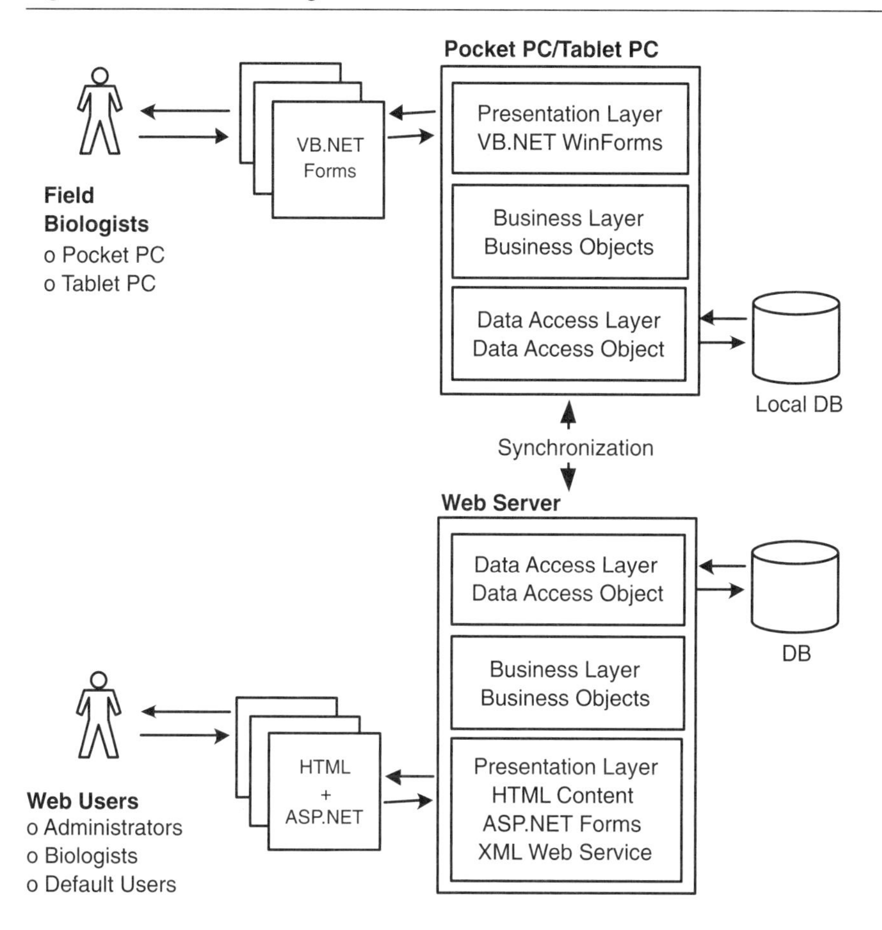

12.2.2 Process Flow

Figure 12–8 illustrates the process flow requirements for the mobile application. Many of these requirements were added for security reasons in order to protect the data on the mobile device in case it is lost or stolen.

When the field biologist first starts the application, the application initially checks to see if the mobile device is connected to the network. The rest of the process depends on the user's network connection state, as described below.

User connected to network. If the user is connected to the network, the application displays the login page.

The user enters his/her username and password on the login page and attempts to log in remotely. If the user login fails, he/she can attempt to log in again. However, if the user fails three times, the application deletes the local data as a security precaution and exits the application.

If the login succeeds, the user may select a task to perform. If the user has made updates locally, then the valid tasks are to "Edit Data Locally" or "Upload Changes." If the user has not made changes, the valid tasks are to "Download Data" or "Edit Data Locally." The application does not allow a user to download new data if there are local changes that have not been uploaded to the server.

The user may decide to log in locally even though a network connection is present by selecting the "Login locally" check box. This will minimize lengthy roundtrips to the

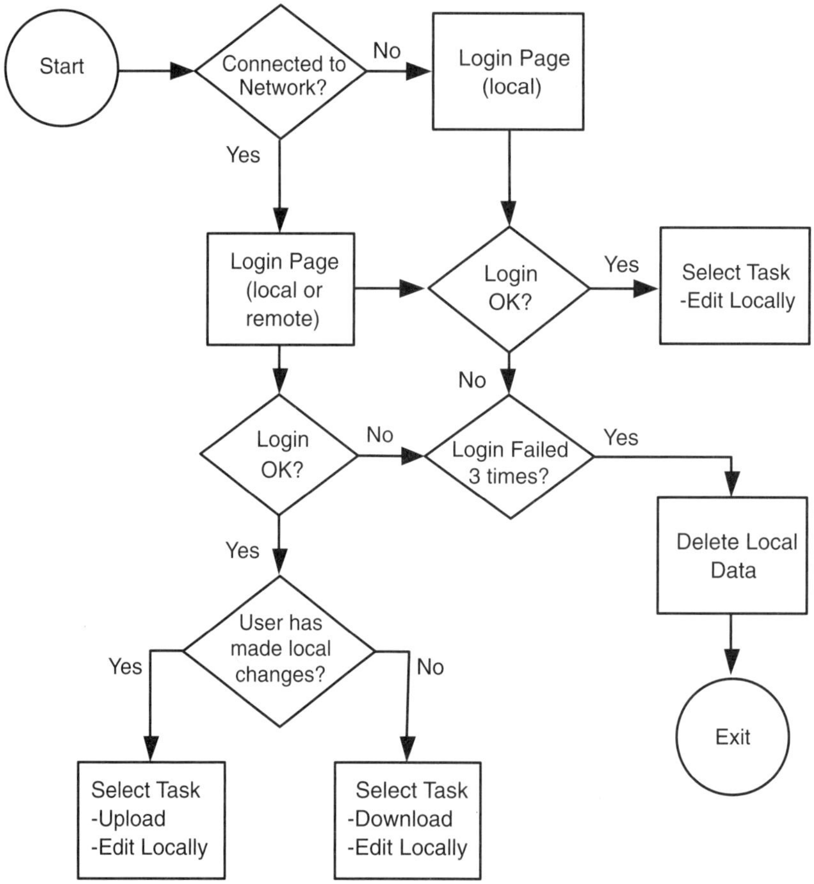

Figure 12–8 Process flow requirements for the mobile application

server if the user only intends to edit local data. If the user logs in locally, the same process flow applies as if the user was disconnected.

User disconnected from network. If the user is not connected to the network, he/she may only log in locally. If the local login fails, the user may try again, up to three times. After three attempts, the application deletes the data from the local database as a security precaution and exits the application. The user will have to reconnect to the network and successfully login remotely to re-initialize the application before he/she will be able to work locally again.

If the local login succeeds, the user may select a task to perform. Currently, the only valid task that the user can perform when disconnected from the network is "Edit Data Locally."

12.3 POCKET PC CLIENT DETAILED DESIGN

This section describes the detailed design of the Pocket PC Mobile Biologist application, including its class diagrams and object models.

All the objects in the Pocket PC client application are in the namespace "MobileBiologist.PPC." As you can see in Figure 12–9, we have created several different sub-namespaces to hold the various objects.

Figure 12–9 Namespaces in the Mobile Biologist Pocket PC application

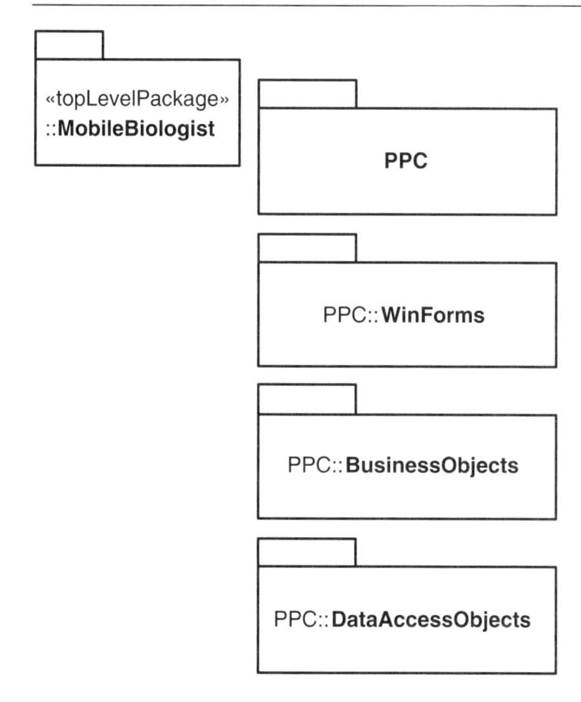

Figure 12–10 Windows Forms in the Pocket PC application

WinForms::**frmLogin**

WinForms::**frmTasks**

WinForms::**frmWaterSource**

WinForms::**frmSampleDetails**

WinForms::**frmDownload**

WinForms::**frmUpload**

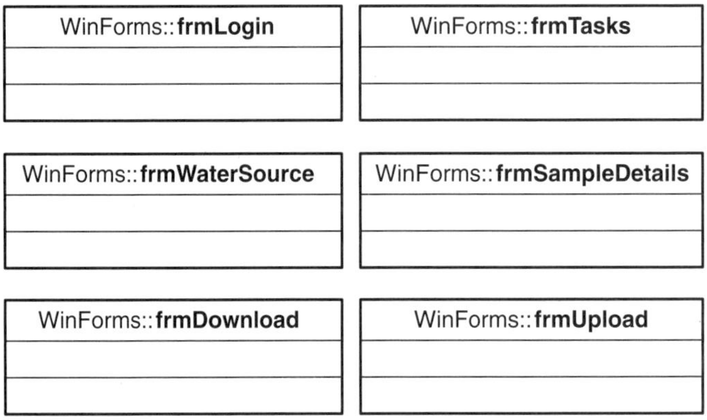

12.3.1 Presentation Layer

The Presentation Layer consists of Windows Forms developed using VB.NET to run on the .NET Compact Framework. We created all of the forms as classes in the MobileBiologist.PPC.WinForms namespace (see Figure 12–10). The rest of this section describes each form in more detail.

1. **Login form.** The login form, frmLogin, allows a user to log in to the Pocket PC application (see Figure 12–11). If the user is connected to the network, he/she may log in locally or remotely. If not, he/she may only log in locally.
2. **Task selection form.** The task selection form, frmTasks, allows the user to choose a task to perform by clicking a button (see Figure 12–12). Some tasks may only be performed when the user is connected to the network and the buttons for these tasks will be disabled if the user logs in locally. In addition, some tasks are dependent on the current state of the local database. For example, the "Upload Changes to Server" button is only enabled if there have been changes made to the local database.

Figure 12–11 Pocket PC login form

Figure 12–12 Pocket PC task selection form

3. **Water source details.** The water source details form, frmWaterSource, shows the details of a water source and a list of the existing corresponding water samples (see Figure 12–13). The user may select a different water source using the drop-down list at the top of the page. In addition, the user may add a new sample by clicking the "Add Sample" button. The user may also select a water sample from the list box and click the "View Details" button to see the sample's detailed information.
4. **Add/view sample form.** The frmSampleDetails form is used to add and/or view a sample (see Figure 12–14). If the user adds a sample, the form's text boxes and drop-down lists are enabled and the user may enter and save data. If the user only wants to view a sample, the same form is used. The text boxes and drop-down lists are populated with the actual data but they are disabled so that they cannot be changed.

Figure 12–13 Pocket PC water source details form

Figure 12–14 Pocket PC add/view sample form

5. **Download form.** The frmDownload form downloads water sources and displays them in a list (see Figure 12–15). The user can then decide whether or not to download the data or cancel the action. Currently, we only support an all-or-nothing download. However, it would be possible to allow the user to select only certain items to download from the list.
6. **Upload form.** The frmUpload form shows new samples the user has added locally and allows the user to upload them to the server (see Figure 12–16). We currently use an all-or-nothing approach where the user must upload all changes at once. However, it would be possible to allow the user to select only certain items to upload.

Figure 12–15 Pocket PC download form

Figure 12–16 Pocket PC upload form

12.3.2 Business Logic Layer

This section describes all the objects in the MobileBiologist.PPC.BusinessObjects namespace.

1. **Water Source class.** The Water Source class in the Pocket PC application represents a water source that a field biologist will visit to take water samples (see Figure 12–17). This class is very similar to the Water Source class that is part of the server application, discussed later in this chapter. Indeed, the application's Water Source class is a stripped-down version of the server class and only contains methods that are used in the Pocket PC application. The Water Source class on the Pocket PC contains the following public properties:

   Id. This is an integer that uniquely identifies the water source.

   Name. This is a descriptive name of the water source (e.g., Central Park Reservoir).

Figure 12–17 Pocket PC Water Source class diagram

BusinessObjects::**WaterSource**
-Id : Integer -Name : String -Location : String -Latitude : String -Longitude : String -Samples : ArrayList
+New() +New(in Id) +GetWaterSourceList() : ArrayList +Insert() : Boolean +GetRemoteWaterSourceXML() : String

Location. This describes the location of the water source (e.g., 90th and 5th Avenue).

Latitude and Longitude. These are coordinates that identify the exact location of the water source. They could be used in conjunction with a GPS device to pinpoint the location where a measurement was taken.

Samples. This is an array list of the samples that have been taken for this water source.

The Water Source class in the Pocket PC application contains the following public methods:

New. This method is the default constructor.

New (Id). This constructor queries the local database to populate all of the fields based on the water source's identifier, which is passed as a parameter.

GetWaterSourceList. This method returns an array list containing all of the water sources found in the local database.

Insert. This method inserts a water source into the local database.

GetRemoteWaterSourceXML. This method calls the web service on the server to retrieve an XML listing of all water sources, including their full details and samples.

2. **Sample class.** The Sample class represents an actual water sample taken by a field biologist (see Figure 12–18). This class is very similar to the Sample class that is part of the server application discussed later in this chapter. Again, this application's Sample class is a modified version of the server class and only contains methods that are used in the Pocket PC application. However, it also contains a few extra methods and properties that

Figure 12–18 Pocket PC Sample class diagram

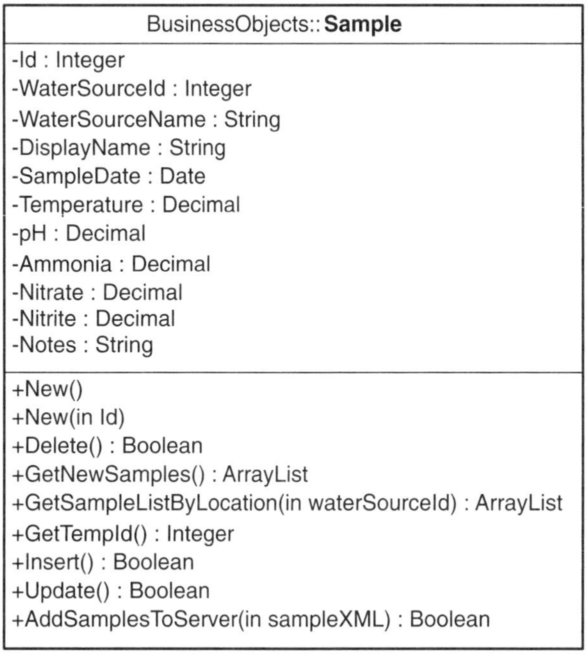

BusinessObjects::**Sample**
-Id : Integer -WaterSourceId : Integer -WaterSourceName : String -DisplayName : String -SampleDate : Date -Temperature : Decimal -pH : Decimal -Ammonia : Decimal -Nitrate : Decimal -Nitrite : Decimal -Notes : String
+New() +New(in Id) +Delete() : Boolean +GetNewSamples() : ArrayList +GetSampleListByLocation(in waterSourceId) : ArrayList +GetTempId() : Integer +Insert() : Boolean +Update() : Boolean +AddSamplesToServer(in sampleXML) : Boolean

are specific to the Pocket PC implementation of the class. The Sample class contains the following public properties:

Id. This is a unique identifier. All water samples downloaded from the server will have an identifier that is a positive integer. All new samples added on the Pocket PC will have an identifier that is a negative integer. We can use this field to determine which samples are new and need to be uploaded. New samples will get a new positive integer Id when they are inserted into the main server database.

WaterSourceId. This is the unique identifier of the water source from which this sample was taken.

WaterSourceName. This is the name of the water source from which this sample was taken. This is read-only and is used for display purposes.

DisplayName. This is a combination of the water source name and the sample date (see below). It is read-only and used for display purposes.

SampleDate. This is the date on which the sample was taken.

Temperature. This is the temperature of the water sample.

pH. This is the pH of the water sample.

Ammonia, Nitrate, and Nitrite. These are the levels of ammonia, nitrate, and nitrite present in the water sample, measured in parts per million.

Notes. This field contains notes about the water sample.

The Sample class in the Pocket PC application contains the following public methods:

New. This method is the default constructor.

New (Id). This constructor queries the local database to populate all of the fields based on the water sample's identifier, which is passed as a parameter.

Delete. This method deletes the sample from the local database.

GetNewSamples. This method returns an array list of all new samples that were added on the Pocket PC (those with negative identifiers).

GetSampleListByLocation. This method returns an array list containing all the samples for a given water source.

GetTempId. This method determines the next negative integer to assign as a temporary identifier to a new sample.

Insert. This method adds the sample to the local database.

Update. This method updates the sample in the local database.

AddSamplesToServer. This method takes an XML string containing all of the samples added locally. It passes this string to a method on the web service that adds all the new samples to the remote database.

3. **Utilities class.** The Utilities class contains a variety of helper functions that are used by the Pocket PC application (see Figure 12–19). This class contains no public properties,

Figure 12–19 Pocket PC Utilities class diagram

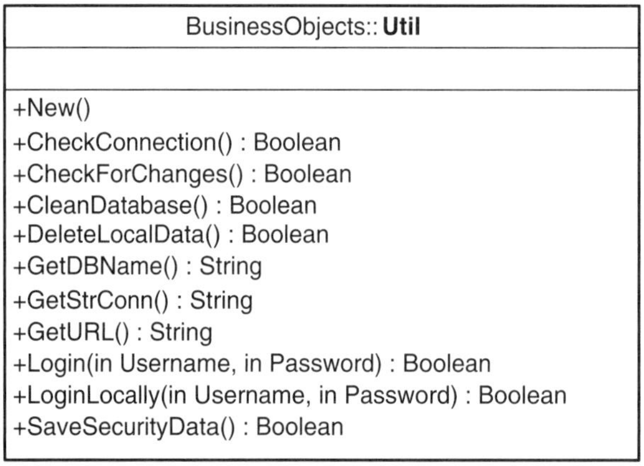

BusinessObjects:: **Util**
+New() +CheckConnection() : Boolean +CheckForChanges() : Boolean +CleanDatabase() : Boolean +DeleteLocalData() : Boolean +GetDBName() : String +GetStrConn() : String +GetURL() : String +Login(in Username, in Password) : Boolean +LoginLocally(in Username, in Password) : Boolean +SaveSecurityData() : Boolean

and it does not have a corresponding class in the server application. It contains the following public methods:

New. This method is the default constructor.

CheckConnection. This method determines whether or not the Pocket PC has an active network connection by calling a method on the web service.

CheckForChanges. This method determines whether or not the user has added samples to the local database by calling the GetNewSamples method on the Sample class.

CleanDatabase. This method deletes all the water sources and samples from the local database in preparation for a download of new data.

DeleteLocalData. This method deletes all data from the local database when a user has failed three login attempts.

GetDBName. This method returns the name of the database.

GetStrConn. This method returns the database connection string.

GetURL. This method returns the URL of the web service. This is found in the file config.xml, located in the same directory as the Pocket PC application.

Login. This method validates the given username and password against the remote web service.

LoginLocally. This method validates the username and password against the local database.

SaveSecurityData. This method saves the username and password in the local database after a user conducts a successful remote login so that the user can log in locally at a later date.

4. **Server Proxy class.** The Server Proxy class communicates with the web service on the remote server (see Figure 12–20). This class is auto-generated by Microsoft Visual Studio .NET when a web reference is added. This class contains a default URL that is set when the class is created. This URL can be overridden after an instance of the class is created. The URL is stored in a configuration XML file that gets deployed with the application. This makes it possible to change the URL without rebuilding the application. This class will only need to change if the web service changes. In that case, the web reference should be updated and the class will be regenerated. This class has no public properties. It has the following public methods:

   New. This constructor sets the default URL of the web service. This property is reset by the application after an instance of the class is created.

   AddAllSamples. This method takes an XML string containing all of the water samples added locally and sends it to the server so that the samples can be added to the remote database.

Figure 12–20 Pocket PC Server Proxy class diagram

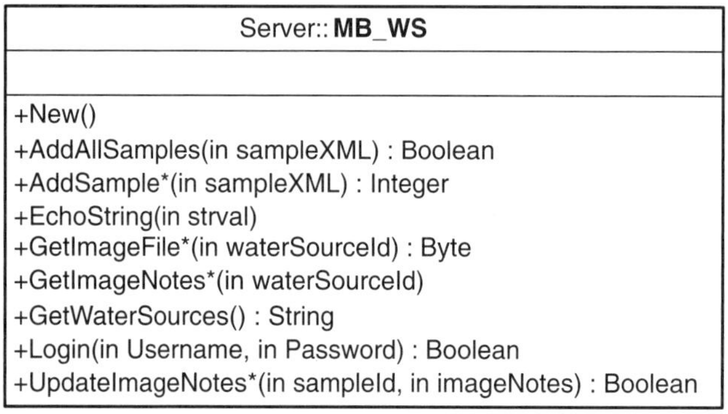

Server:: **MB_WS**
+New() +AddAllSamples(in sampleXML) : Boolean +AddSample*(in sampleXML) : Integer +EchoString(in strval) +GetImageFile*(in waterSourceId) : Byte +GetImageNotes*(in waterSourceId) +GetWaterSources() : String +Login(in Username, in Password) : Boolean +UpdateImageNotes*(in sampleId, in imageNotes) : Boolean

* Only used by Tablet PC application

AddSample. This method is used to add one sample at a time. It is only used by the Tablet PC application, discussed later in this chapter.

EchoString. This method sends a string to the remote server, which returns a date. It is used to test the web service and see if a connection can be established with the server.

GetImageFile. This method returns an image of the water source. It is only used by the Tablet PC application, discussed later in this chapter.

GetImageNotes. This method returns the markings that were added to the water source image, indicating where the sample was taken. It is only used by the Tablet PC application, discussed later in this chapter.

GetWaterSources. This method retrieves an XML listing of water sources from the remote server.

Login. This method validates the username and password against the remote server.

UpdateImageNotes. This method saves the markings that were added to the water source image, indicating where the sample was taken. It is only used by the Tablet PC application, discussed later in this chapter.

Ideally, you should set up SSL on your web server so that all of the calls to the web service can be made using HTTPS and all of the data sent will be encrypted. However, for the purposes of this sample application, we did not do so.

12.3.3 Data Access Layer

The client database uses Microsoft SQL Server CE 2.0, which is secured through the use of a password. For simplicity, we decided not to encrypt the entire database, although this could be done by installing the Microsoft High Encryption Pack for Pocket PC v1.0 and modifying the database connection string to specify that encryption is used.

All communication with the database is carried out through an object in the MobileBiologist.PPC.DataAccessObjects namespace. This section describes that object and the actual database schema.

1. **SQLCEHelper Class.** The SQLCEHelper class provides a set of functions for manipulating the local database and performing queries (see Figure 12–21). It is a scaled-down version of the SQLHelper class, which is used by the server application. The SQLCEHelper class has the following public methods:

    New. This method is the default constructor.

    CreateDB. This method creates a database with the connection string passed as a parameter. This connection string includes the database name and password.

    DBExists. This method checks to see if the specified database exists.

    DeleteDB. This method deletes the specified database.

    ExecuteCommand. This method connects to the specified database and executes the given SQL statement.

    ExecuteDataset. This method executes the given SQL statement and returns a dataset.

    ExecuteScalar. This method executes the given SQL statement and returns the scalar value returned by the SQL call.

Figure 12–21 Pocket PC SQLCEHelper class diagram

DataAccessObjects::**SQLCEHelper**
+New() +CreateDB(in DBName) : Boolean +DBExists(in DBName) : Boolean +DeleteDB(in DBName) : Boolean +ExecuteCommand(in strConn, in strSQL) +ExecuteDataset(in strConn, in strSQL) : DataSet +ExecuteScalar(in strConn, in strSQL) : Integer

2. **Database Schema.** The local database contains three tables:

    1. Security
    2. WaterSource
    3. Sample

    The Security table contains the following fields:

Field Name	Data Type	Primary Key
Username	nvarchar(50)	Yes
Pwd	nvarchar(50)	

The Water Source table contains the following fields:

Field Name	Data Type	Primary Key
Id	int	Yes
Name	nvarchar(255)	
Location	nvarchar(255)	
Latitude	nvarchar(50)	
Longitude	nvarchar(50)	

The Sample table contains the following fields:

Field Name	Data Type	Primary Key
Id	int	Yes
WaterSourceId	int	
SampleDate	nvarchar(50)	
Temperature	float	
pH	float	
Notes	nvarchar(255)	
Ammonia	float	
Nitrate	float	
Nitrite	float	

12.4 TABLET PC CLIENT DETAILED DESIGN

This section describes the detailed design of the Tablet PC Mobile Biologist application, including its class diagrams and object models.

All the objects in the Tablet PC client application are in the namespace "MobileBiologist.Tablet." As you can see in Figure 12–22, we have created several different sub-namespaces to hold the various objects.

Figure 12–22 Namespaces in the Mobile Biologist Tablet PC application

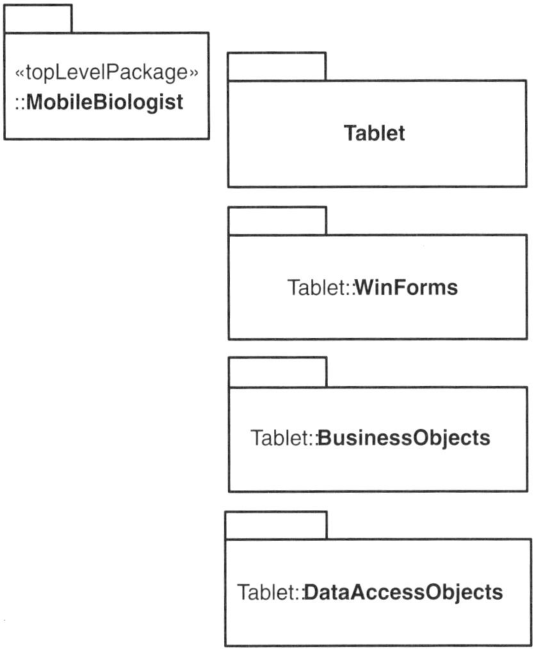

«topLevelPackage»
::**MobileBiologist**

Tablet

Tablet::**WinForms**

Tablet::**BusinessObjects**

Tablet::**DataAccessObjects**

12.4.1 Presentation Layer

The Presentation Layer consists of Windows Forms developed using Microsoft VB.NET to run on Microsoft Windows XP Tablet Edition. To enhance the user experience on the Tablet PC, the forms use the InkEdit control instead of text boxes wherever possible. This text recognition control allows field biologists to enter handwritten information using an electronic pen. (The InkEdit control also allows users to type in information so it will also work on a Laptop PC.)

The InkEdit control and InkPicture control (mentioned later in this section) come with the Microsoft Tablet PC SDK. You must follow a few steps to use these controls in your application. For example, in your VB.NET project, you must initially add a reference to the Microsoft.Ink Library. You must then customize the toolbox to add these controls. You can do this by right-clicking on the toolbox and selecting Add/Remove Items. You then select the controls you want to use in your application.

We created all the application forms as classes in the MobileBiologist.Tablet.WinForms namespace (see Figure 12–23). Each of the forms inherits from a parent class called frmTemplate. The rest of this section describes each form in more detail.

1. **Template form.** All forms in the Tablet PC application inherit from a base class called frmTemplate. This class defines the UI elements that are common to all the forms, such

Figure 12–23 Windows Forms in the Tablet PC application

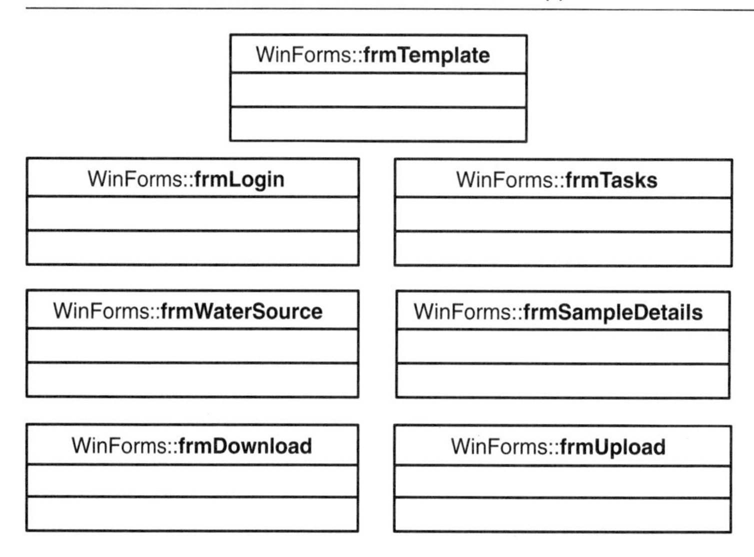

as the File menu and the top banner. By using form inheritance, we can change the look and feel of the template form and the changes will be seen in all of the inherited forms. We were not able to use forms inheritance in the Pocket PC application because this feature is not available. However, the Tablet PC application is a normal Windows Forms application running on the .NET Framework, where visual inheritance is supported.

2. **Login form.** The login form, frmLogin, allows the user to log in to the Tablet PC application (see Figure 12–24). If the user is connected to the network, he/she may log in locally or remotely. If not, the user may only log in locally.
3. **Task selection form.** The task selection form, frmTasks, allows the user to choose a task to perform by clicking a button (see Figure 12–25). Some tasks may only be performed when the user is connected to the network, so the buttons for these tasks will be disabled if the user logs in locally. In addition, some tasks are dependent on the current state of the local database. For example, the "Upload Changes to Server" button is only enabled if there have been changes made to the local database.
4. **Water Source Details.** The water source details form, frmWaterSource, shows the details of a water source and a list of the existing corresponding water samples (see Figure 12–26). The user may select a different water source using the drop-down list at the top of the page. The user may add a new sample by clicking the "Add Sample" button. The user also may select a water sample from the list box and click the "View Details" button to see the sample's detailed information.

Figure 12–24 Tablet PC login form

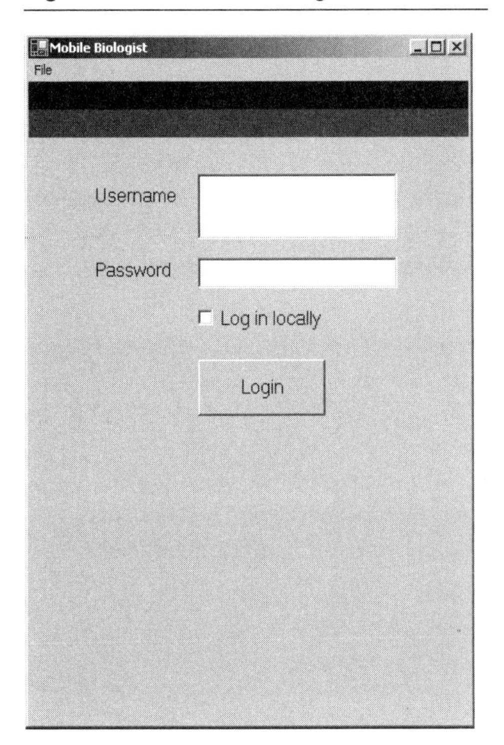

Figure 12–25 Tablet PC task selection form

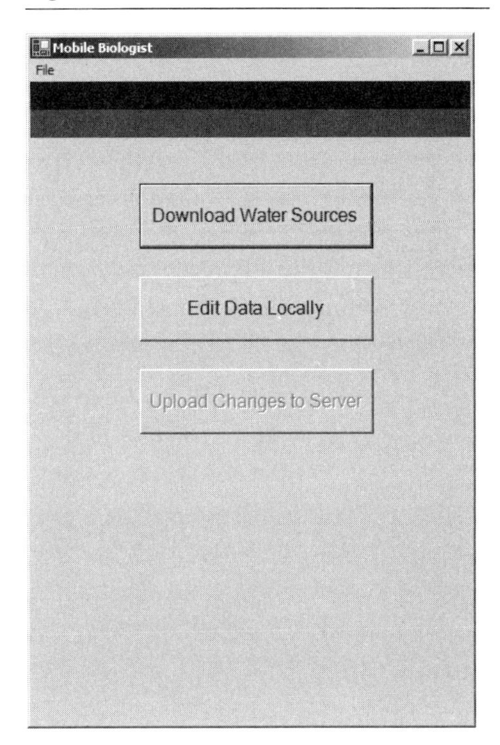

5. **Add/view sample form.** The frmSampleDetails form is used to add and/or view a sample (see Figure 12–27). If the user adds a sample, the form's text boxes and drop-down lists are enabled and the user may enter and save data. If the user wants to view a sample, the same form is used. The text boxes and drop-down lists are populated with the actual data but they are disabled so that they cannot be changed.

   This form uses the Microsoft Tablet PC SDK InkPicture control to display the image of the sample's water source which is stored in the database. (These images are initially uploaded through the administrative web site.) A field biologist can use the Ink capabilities of the Tablet PC to mark the image indicating where the water sample was obtained. This Ink is then saved in the database as an array of bytes.

   When the user views an existing sample, the image and Ink markings are retrieved from the database. First, the image is written to a temporary file that is used as the image source for the InkPicture control. The array of bytes that represents the markings is then loaded into a new Ink object which is used for the Ink property of the InkPicture control. The control is then refreshed.

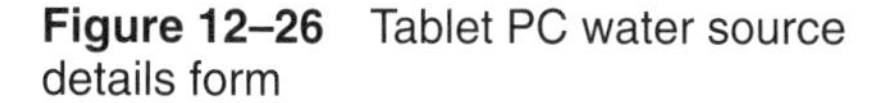

Figure 12–26 Tablet PC water source details form

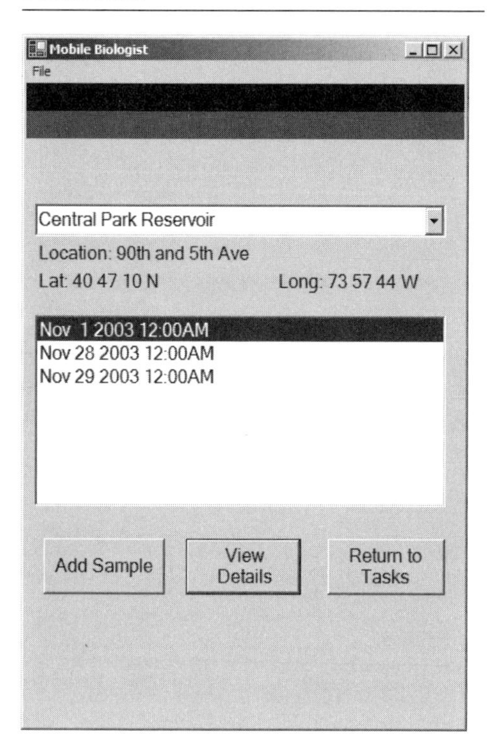

Figure 12–27 Tablet PC add/view sample form

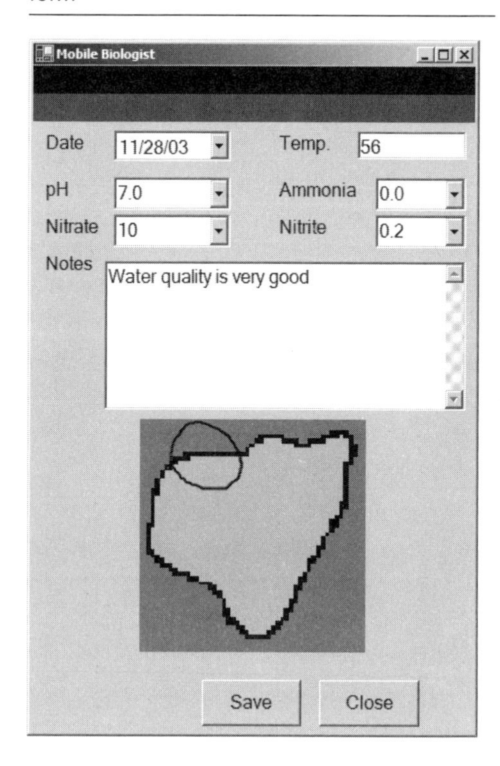

It is important to note that these markings can only be viewed in the Tablet PC version of the application. Only the Tablet PC's InkPicture control can take an image and overlay this type of marking. This feature, therefore, will not work on the Pocket PC.

6. **Download form.** The frmDownload form downloads water sources and displays them in a list (see Figure 12–28). The user can then decide whether or not to download the data or cancel the action. At this point, we only support an all-or-nothing download. However, it would be possible to allow the user to select only certain items to download from the list.
7. **Upload form.** The frmUpload form shows a list of new samples the user has added locally and allows the user to upload these samples to the server (see Figure 12–29). Again, we currently use an all-or-nothing approach, where the user must upload all changes at once. However, it would be possible to allow the user to select only certain items to upload.

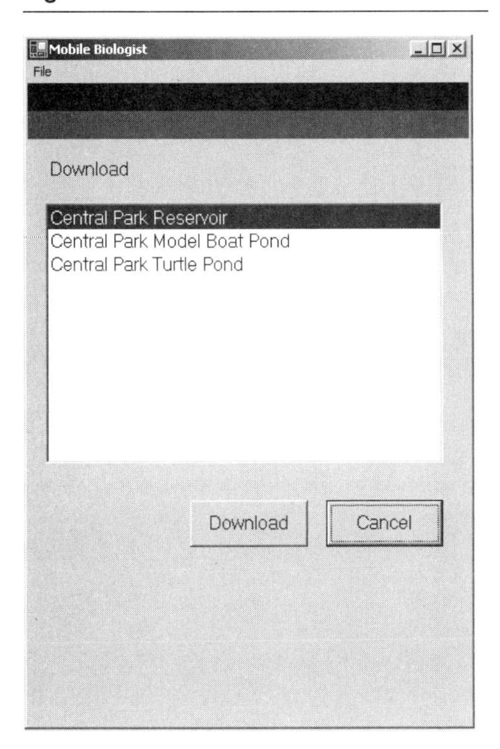

Figure 12–28 Tablet PC download form

Figure 12–29 Tablet PC upload form

12.4.2 Business Logic Layer

This section describes the objects in the MobileBiologist.Tablet.BusinessObjects namespace.

1. **Water Source class.** The Water Source class in the Tablet PC application represents a water source that a field biologist will visit to take water samples (see Figure 12–30). This class is very similar to the Water Source class that is part of the server application, discussed later in this chapter. Indeed, the application's Water Source class is a stripped-down version of the server class and only contains methods that are used in the Tablet PC application. The Water Source class on the Tablet PC contains the following public properties:

   Id. This is an integer that uniquely identifies the water source.

   Name. This is a descriptive name of the water source (e.g., Central Park Reservoir).

   Location. This describes the location of the water source (e.g., 90th and 5th Avenue).

Figure 12–30 Tablet PC Water Source class diagram

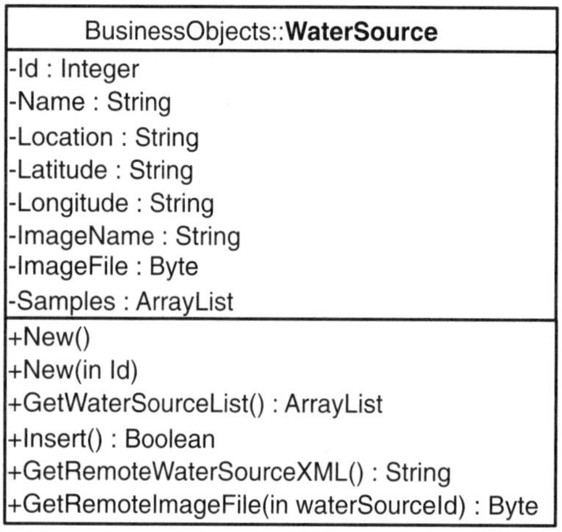

BusinessObjects::**WaterSource**
-Id : Integer -Name : String -Location : String -Latitude : String -Longitude : String -ImageName : String -ImageFile : Byte -Samples : ArrayList
+New() +New(in Id) +GetWaterSourceList() : ArrayList +Insert() : Boolean +GetRemoteWaterSourceXML() : String +GetRemoteImageFile(in waterSourceId) : Byte

Latitude and Longitude. These are coordinates that identify the exact location of the water source. They could be used in conjunction with a GPS device to pinpoint the location where a measurement was taken.

ImageName. This is the name of the water source's image.

ImageFile. This is an array of bytes containing the image of the water source.

Samples. This is an array list of the samples that have been taken for this water source.

The Water Source class in the Tablet PC application contains the following public methods:

New. This method is the default constructor.

New (Id). This constructor queries the local database to populate all of the fields based on the water source's identifier, which is passed as a parameter.

GetWaterSourceList. This method returns an array list containing all of the water sources found in the local database.

Insert. This method inserts a water source into the local database.

GetRemoteWaterSourceXML. This method calls the web service on the server to retrieve an XML listing of all water sources, including their full details and samples.

GetRemoteImageFile. This method returns the array of bytes containing the image file for the water source; this information is not passed via the XML string.

2. **Sample class.** The Sample class in the Tablet PC application represents an actual water sample taken by a field biologist (see Figure 12–31). This class is very similar to the Sample class that is part of the server application, discussed later in this chapter. However, this application's Sample class also contains a few extra methods and properties that are specific to the Tablet PC implementation of the class. The Sample class contains the following public properties:

    Id. This is a unique identifier. All water samples downloaded from the server will have an identifier that is a positive integer. All new samples added on the Tablet PC will have an identifier that is a negative integer. We can use this field to determine which samples are new and need to be uploaded. New samples will get a new positive integer Id when they are inserted into the main server database.

    WaterSourceId. This is the unique identifier of the water source from which this sample was taken.

Figure 12–31 Tablet PC Sample class diagram

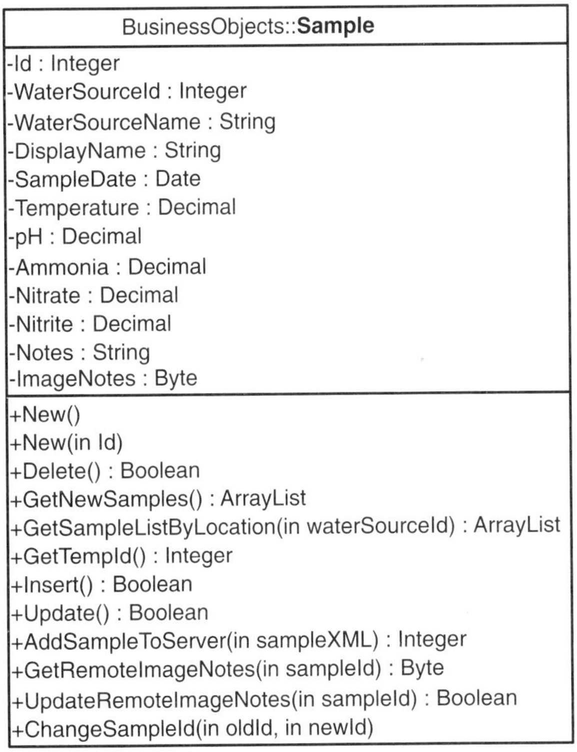

BusinessObjects::**Sample**
-Id : Integer -WaterSourceId : Integer -WaterSourceName : String -DisplayName : String -SampleDate : Date -Temperature : Decimal -pH : Decimal -Ammonia : Decimal -Nitrate : Decimal -Nitrite : Decimal -Notes : String -ImageNotes : Byte
+New() +New(in Id) +Delete() : Boolean +GetNewSamples() : ArrayList +GetSampleListByLocation(in waterSourceId) : ArrayList +GetTempId() : Integer +Insert() : Boolean +Update() : Boolean +AddSampleToServer(in sampleXML) : Integer +GetRemoteImageNotes(in sampleId) : Byte +UpdateRemoteImageNotes(in sampleId) : Boolean +ChangeSampleId(in oldId, in newId)

WaterSourceName. This is the name of the water source from which this sample was taken. This is read-only and is used for display purposes.

DisplayName. This is a combination of the water source name and the sample date (see below). It is read-only and used for display purposes.

SampleDate. This is the date on which the sample was taken.

Temperature. This is the temperature of the water sample.

pH. This is the pH of the water sample.

Ammonia, Nitrate, and Nitrite. These are the levels of ammonia, nitrate, and nitrite present in the water sample, measured in parts per million.

Notes. This field contains notes about the water sample.

ImageNotes. This field contains an array of bytes representing the Ink markings that were made on the water source image, indicating where the sample was taken.

The Sample class in the Tablet PC application contains the following public methods:

New. This method is the default constructor.

New (Id). This constructor queries the local database to populate all of the fields based on the sample's identifier, which is passed as a parameter.

Delete. This method deletes the sample from the local database.

GetNewSamples. This method returns an array list of all new samples that were added on the Tablet PC (those with negative identifiers).

GetSampleListByLocation. This method returns an array list containing all the samples for a given water source.

GetTempId. This method determines the next negative integer to assign as a temporary identifier to a new sample.

Insert. This method adds the sample to the local database.

Update. This method updates the sample in the local database.

AddSamplesToServer. This method takes an XML string containing sample data. It passes this string to a method on the web service that adds the new sample to the remote database. The new Id of the sample is returned.

GetRemoteImageNotes. This method calls the web service to retrieve the array of bytes containing the Ink markings added to the water source image to indicate where the sample was taken. (This information is not included in the XML string retrieved with the list of water sources.)

UpdateRemoteImageNotes. This method sends the updated Ink markings to the web service for storage in the remote database.

ChangeSampleId. This method changes a sample's unique identifier once it has been saved to the remote database via the web service. All new samples initially have a negative integer identifier. Once they are saved on the server and given a positive identifier, this field is updated in the Tablet PC database.

3. **Utilities class.** The Utilities class contains a variety of helper functions that are used by the Tablet PC application (see Figure 12–32). This class contains no public properties and does not have a corresponding class in the server application. It contains the following public methods:

New. This method is the default constructor.

CheckConnection. This method determines whether or not the Tablet PC has an active network connection by calling a method on the web service.

CheckForChanges. This method determines whether or not the user has added samples to the local database by calling the GetNewSamples method on the Sample class.

CleanDatabase. This method deletes all water sources and samples from the local database in preparation for a download of new data.

DeleteLocalData. This method deletes all data from the local database. This is performed when a user has failed three login attempts.

GetStrConn. This method returns the database connection string.

GetURL. This method returns the URL of the web service. This is found in the App.config file, located in the same directory as the Tablet PC application.

Login. This method validates the username and password against the remote web service.

Figure 12–32 Tablet PC Utilities class diagram

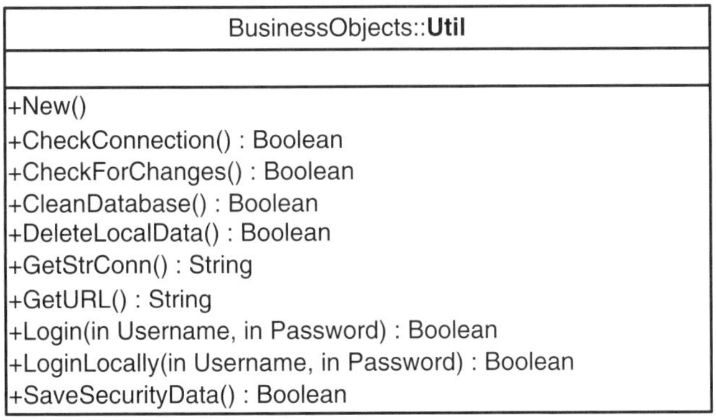

BusinessObjects::**Util**
+New() +CheckConnection() : Boolean +CheckForChanges() : Boolean +CleanDatabase() : Boolean +DeleteLocalData() : Boolean +GetStrConn() : String +GetURL() : String +Login(in Username, in Password) : Boolean +LoginLocally(in Username, in Password) : Boolean +SaveSecurityData() : Boolean

LoginLocally. This method validates the username and password against the local database.

SaveSecurityData. This method saves the username and password in the local database after a user conducts a successful remote login so that the user can log in locally at a later date.

4. **Server Proxy Class.** The Server Proxy class communicates with the web service on the remote server (see Figure 12–33). This class is auto-generated by Microsoft Visual Studio .NET when a web reference is added. This class contains a default URL that is set when the class is created. This URL can be overridden after an instance of the class is created. The URL is stored in the App.config file, which gets deployed with the application. This makes it possible to change the URL without rebuilding the application. This class will only need to change if the web service changes. In that case, the web reference should be updated and the class will be regenerated. This class has no public properties. It has the following public methods:

    New. This constructor sets the default URL of the web service. This property is reset by the application after an instance of the class is created.

    AddAllSamples. This method takes an XML string containing all of the water samples added locally and sends it to the server so that the samples can be added to the remote database.

    AddSample. This method takes an XML string containing one sample's data and sends it to the server so that it can be added to the remote database.

    EchoString. This method sends a string to the remote server, which returns a date. It is used to test the web service and see if a connection can be established with the server.

Figure 12–33 Tablet PC Server Proxy class diagram

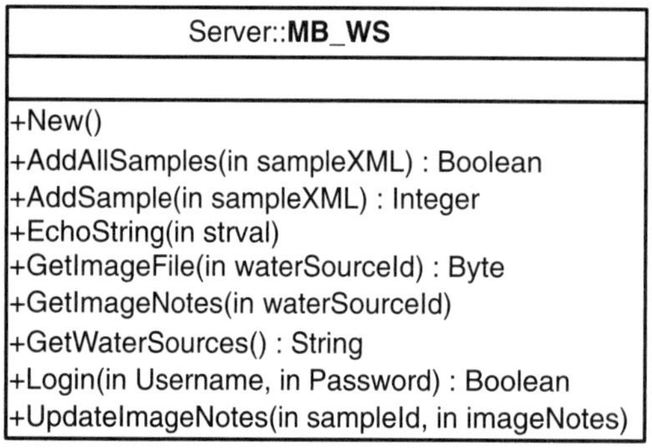

Server::**MB_WS**
+New() +AddAllSamples(in sampleXML) : Boolean +AddSample(in sampleXML) : Integer +EchoString(in strval) +GetImageFile(in waterSourceId) : Byte +GetImageNotes(in waterSourceId) +GetWaterSources() : String +Login(in Username, in Password) : Boolean +UpdateImageNotes(in sampleId, in imageNotes)

GetImageFile. This method returns an array of bytes containing the image of a water source.

GetImageNotes. This method returns an array of bytes containing the Ink markings that were added to the water source image, indicating where the sample was taken.

GetWaterSources. This method retrieves an XML listing of water sources from the remote server.

Login. This method validates the username and password against the remote server.

UpdateImageNotes. This method updates the array of bytes containing the Ink markings that were added to the water source image, indicating where the sample was taken.

Ideally, you should set up SSL on your web server so that all the calls to the web service can be made using HTTPS and all of the data sent will be encrypted. However, for the purposes of this sample application, we did not do so.

12.4.3 Data Access Layer

All communication with the database is carried out through objects in the MobileBiologist.Tablet. DataAccessObjects namespace (see Figure 12–34). This section describes those objects and the actual database schema.

1. **Data Access Object.** The Tablet PC application uses Microsoft SQL Server 2000. As a result, we were able to reuse the data access objects from our server application, which is discussed later. The SQLHelper and SQLParameterCache classes are part of the Data Access Application Block for .NET and were provided by Microsoft for general use.

Figure 12–34 Tablet PC application data access objects

DataAccessObjects::**SQLHelper**

DataAccessObjects::**SQLParameterCache**

2. **Database schema.** The local database contains three tables:

    1. Security
    2. WaterSource
    3. Sample

    The Security table contains the following fields:

Field Name	Data Type	Primary Key
Username	varchar(50)	Yes
Pwd	varchar(50)	

The Water Source table contains the following fields:

Field Name	Data Type	Primary Key
Id	int	Yes
Name	varchar(255)	
Location	varchar(255)	
Latitude	varchar(50)	
Longitude	varchar(50)	
ImageName	varchar(255)	
ImageFile	image	

The Sample table contains the following fields:

Field Name	Data Type	Primary Key
Id	int	Yes
WaterSourceId	int	
SampleDate	smalldatetime	
Temperature	float	
pH	float	
Ammonia	float	
Nitrate	float	
Nitrite	float	
Notes	nvarchar(255)	
ImageNotes	image	

The database contains the following stored procedures:

Procedure Name	Parameters	Return Value
sp_AddSample	@Id int, @WaterSourceId int, @SampleDate smalldatetime, @Temperature float, @pH float, @Ammonia float, @Nitrate float, @Nitrite float, @Notes nvarchar(255), @ImageNotes Image=null	
sp_AddWaterSource	@Id int, @Name varchar(255), @Location varchar(255), @Latitude varchar(50), @Longitude varchar(50), @ImageName varchar(255)=null, @ImageFile image=null	

12.5 SERVER DETAILED DESIGN

All the server objects for the Mobile Biologist application are in the namespace "Mobile Biologist." We have created several different sub-namespaces to hold the various objects. Figure 12–35 illustrates the namespaces in the server application.

12.5.1 XML Web Service

The XML Web Service provides a mechanism for the mobile device to communicate with the server. We decided to use a web service to handle the data synchronization between the client and the server rather than RDA or Merge Replication. One reason for this is that we had already developed a set of business objects for use by the server web site and it was very simple to expose these objects as a web service using SOAP. In addition, we wanted all communication with the server database to go through the server business objects so that they can perform any necessary validation.

There is only one class in this namespace, namely MB_WS (see Figure 12–36). The class MB_WS stands for "Mobile Biologist Web Service." It includes the following public methods:

New. This method is the default constructor.

AddAllSamples. This method takes an XML string that contains the new samples from the client application. It iterates through the XML document and creates new Sample business objects, populating them with the XML data. It then saves each object to the database.

AddSample. This method takes an XML string that contains the details for one new sample from the client application. It saves the new Sample in the database and returns the new identifier to the client application. This method is only used by the Tablet PC application.

Figure 12–35 Namespaces in the Mobile Biologist server application

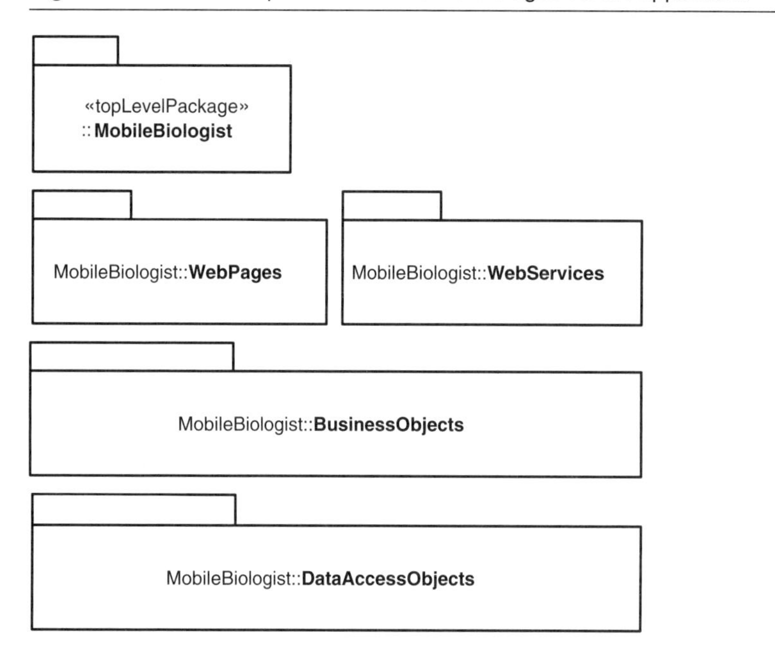

Figure 12–36 Server MB_WS class diagram

WebServices::**MB_WS**
+New() +AddAllSamples(in sampleXML) : Boolean +AddSample*(in sampleXML) : Integer +EchoString(in strval) : Date +GetImageFile*(in waterSourceId) : Byte +GetImageNotes*(in waterSourceId) : Byte +GetWaterSources() : String +Login(in Username, in Password) : Boolean +UpdateImageNotes*(in sampleId, in imageNotes) : Boolean

* Only used by Tablet PC application

EchoString. This is a simple method that returns a date; it is used to test connectivity to the web service.

GetImageFile. This method returns an array of bytes containing an image of the specified water source. This method is necessary because the array of bytes is not included in the XML string returned by the GetWaterSources method (see below). It is only used by the Tablet PC application.

GetImageNotes. This method returns an array of bytes containing the Ink markings that were made on the water source image by the Tablet PC application. This method is necessary because the array of bytes is not included in the XML string returned by the GetWaterSources method (see below). It is only used by the Tablet PC application.

GetWaterSources. This method returns an XML string with all the water sources and their sample data. It retrieves an array list of water sources from the Water Source business object. It then iterates through each item in the list, reads the properties, and adds the data to the XML file.

Login. This method takes a username and password, and validates them against the database.

UpdateImageNotes. This method accepts the array of bytes containing the Ink markings for a given water sample indicating where that sample was taken, and updates the data in the database. This method is necessary because this information is not included in the Sample XML data accepted by the AddSample method. It is only used by the Tablet PC application.

12.5.2 Presentation Layer

The presentation layer consists of a set of HTML and ASP.NET web pages and user controls, as described below.

1. **Utility web pages and controls.** Figure 12–37 illustrates some of the utility web pages and controls. The utility web pages include:

Figure 12–37 Utility web pages and controls

WebPages::**Global**

WebPages::**pagetemplate**

WebPages::**Calendar**

WebPages::**Banner**

WebPages::**LeftNav**

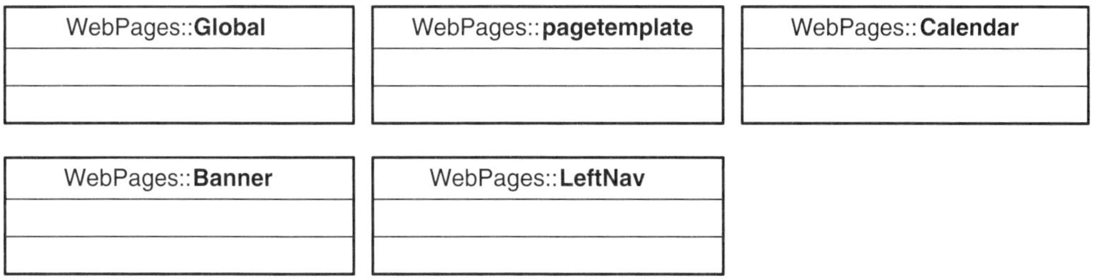

Global. This page replaces the traditional Global.asa file in ASP applications. The Global class provides event handlers such as Session_Start, Application_Start, Application_AuthenticateRequest, etc. The Application_AuthenticateRequest method creates a new Custom Principal object and adds it to the current HTTP Context.

Pagetemplate. This is an ASP.NET page that we used to develop the user interface for the web site. Every time we create a new ASP.NET page, we use this page as a template, cutting and pasting its HTML into the new page.

Calendar. This pop-up page allows users to select a date.

Banner. This control displays a banner at the top of the page.

LeftNav. This control displays the left-hand navigation bar. It checks the role of the logged-in user to decide which links to display.

2. **Default web pages.** Figure 12–38 illustrates the default web pages that are available to all users. These include:

Login. The web site is protected by Microsoft Forms Authentication. Each time an unauthenticated user visits the web site, he/she will be automatically redirected to this page. This is done by modifying the web site's Web.config file as shown:

```
<authentication mode="Forms">
<forms loginUrl="/MB_WebSite/login.aspx"
        name="sqlAuthCookie" timeout="60" path="/">
</forms>
</authentication>
<authorization>
<deny users="?"/>
<allow users="*" /> <!– Allow all authenticated users –>
</authorization>
```

Figure 12–38 Default web pages

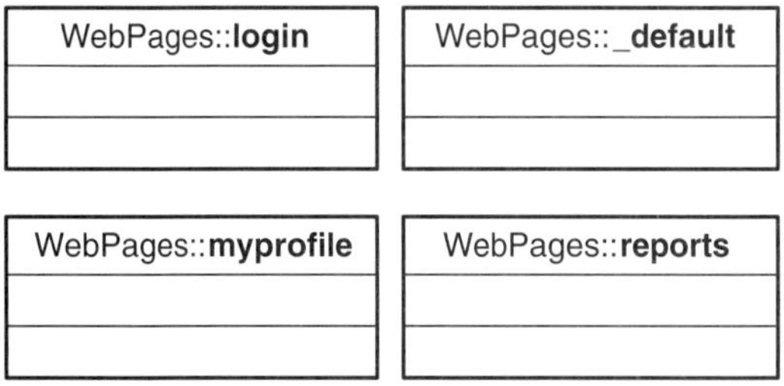

WebPages::**login**

WebPages::**_default**

WebPages::**myprofile**

WebPages::**reports**

The login page allows users to enter a username and password. If the entries are valid, the page issues the Forms Authentication token. It also sets a cookie containing user information. This cookie is used by the Global.asa's Application_AuthenticateRequest method to create the Custom Principal object. The Custom Principal object is used by the Forms Authentication filter to perform role-based security.

This page is also used to register new users by passing method=register in the query string. Users who wish to register may enter their username, password, first name, last name, and email address. The application automatically adds newly registered users into the Default user role. The user must contact an administrator in order to be assigned a role other than the Default user role.

This page is also used to log users out of the web site by passing method=logout in the query string. The page logs the user out of the web site by calling the FormsAuthentication.SignOut method. It also deletes the user information cookie and removes the user from the HTTP Context.

Default. This is the home page of the site.

MyProfile. This page allows each user to edit his/her profile information, including name, email, and password.

Reports. This page allows users to view water quality reports.

3. **Biologist web pages.** There are several pages that are only accessible to biologists and administrators (see Figure 12–39). These pages are used to add, update, and delete water source and sample data.

   To secure these pages, we placed them in a subdirectory called "updatedata." We also placed a Web.config file in that directory, which indicates that the pages are only accessible to users with an "ADMIN" or "BIOLOGIST" role.

Figure 12–39 Biologist web pages

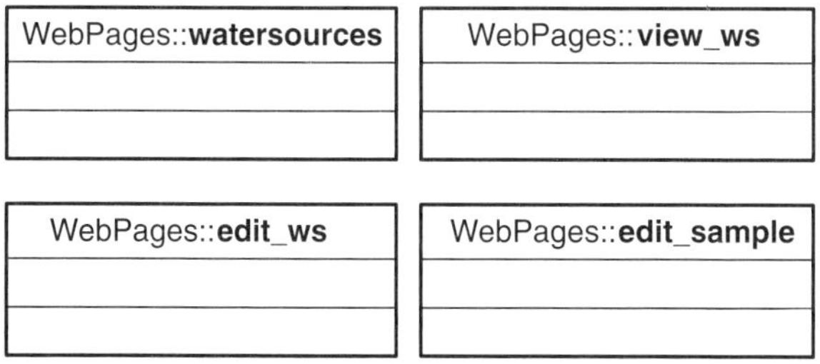

WebPages::**watersources**

WebPages::**view_ws**

WebPages::**edit_ws**

WebPages::**edit_sample**

```
<authorization>
<deny users="?"/>
<allow roles="ADMIN, BIOLOGIST"/>
<deny roles="DEFAULT"/>
</authorization>
```

The web pages that are accessible to both biologists and administrators include:

WaterSources. This page shows a list of water sources from the database.

View_WS. The "View Water Source" page shows the details of a water source and its samples.

Edit_WS. The "Edit Water Source" page allows a biologist to edit a water source's details, such as name and location.

EditSample. This page allows a biologist to edit a water sample. This same page is used to add a new sample to the database.

4. **Administrative web pages.** There are several pages that are only accessible to administrators (see Figure 12–40). These pages are used to add, update, and delete users and roles. To secure these pages, we placed them in a subdirectory called "admin." We also placed a Web.config file in that directory, which indicates that the pages are only accessible to users with an "ADMIN" role.

```
<authorization>
<deny users="?"/>
<allow roles="ADMIN"/>
<deny roles="DEFAULT, BIOLOGIST "/>
</authorization>
```

Figure 12–40 Administrator web pages

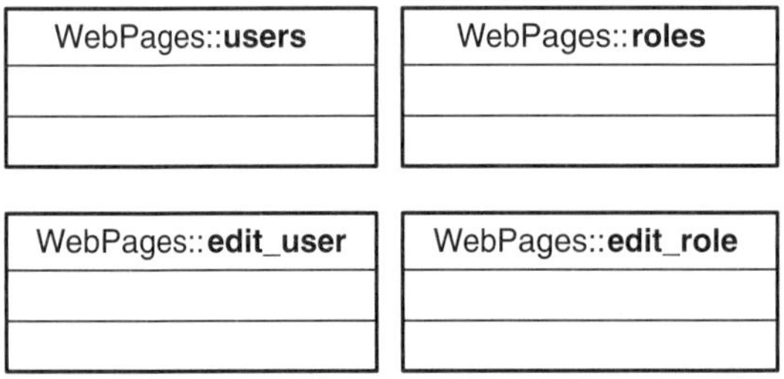

WebPages::**users**	WebPages::**roles**
WebPages::**edit_user**	WebPages::**edit_role**

The web pages that are only accessible to administrators include:

Users. This page allows an administrator to view a list of users.

Roles. This page allows an administrator to view a list of roles.

EditUser. This page allows an administrator to edit a user's information, including what role the user belongs to. The same page is also used to add a new user.

EditRole. This page allows an administrator to edit role information, such as name and description. The same page is also used to add a new role.

12.5.3 Business Logic Layer

This section describes all objects in the MobileBiologist.BusinessObjects namespace.

1. **MBUser class.** The MBUser class represents a user of the Mobile Biologist application (see Figure 12–41). It has the following public properties:

   Id. This is an integer that uniquely identifies the user.

   FirstName. This is the user's first name.

   LastName. This is the user's last name.

Figure 12–41 Server MBUser class diagram

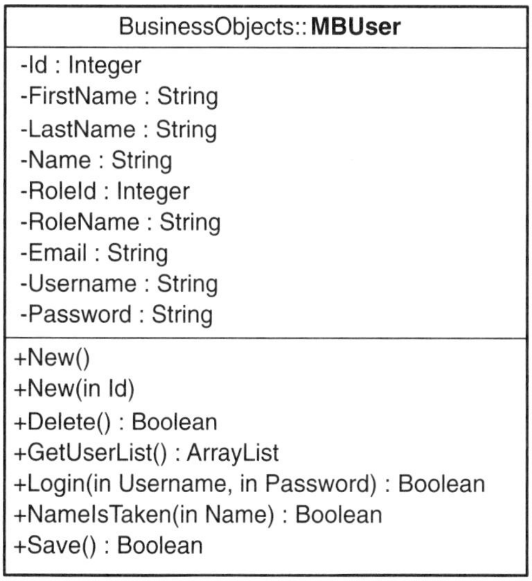

BusinessObjects:: **MBUser**
-Id : Integer -FirstName : String -LastName : String -Name : String -RoleId : Integer -RoleName : String -Email : String -Username : String -Password : String
+New() +New(in Id) +Delete() : Boolean +GetUserList() : ArrayList +Login(in Username, in Password) : Boolean +NameIsTaken(in Name) : Boolean +Save() : Boolean

Name. This is a combination of the user's first and last name. It is read-only and used for display purposes.

RoleId. This is the unique identifier of the role the user belongs to.

RoleName. This is the name of the role the user belongs to.

Email. This is the user's email address.

Username. This is the username used to log in to the web site and the mobile application.

Password. This is the user's password. For simplicity, we did not encrypt it in this sample application. However, it could be encrypted using the Forms Authentication object's HashPasswordForStoringInConfigFile method.

The MBUser class has the following public methods:

New. This method is the default constructor.

New (Id). This constructor queries the database to populate all of the fields based on the user's identifier, which is passed as a parameter.

Delete. This method deletes the user from the database.

GetUserList. This method returns an array list of MBUser objects, representing every user in the database.

Login. This method verifies the username and password against the value in the database.

NameIsTaken. This method is used to check if a particular username has already been taken. It is useful for preventing duplicates whenever you are adding a new user or changing an existing user's username.

Save. This method saves the user's data to the database. If the user Id is zero, the method inserts the record, retrieves the new user Id, and populates it. If the user Id is greater than zero, the method performs an update.

2. **MBRole class.** The MBRole class represents a security role for the Mobile Biologist web application (see Figure 12–42). It has the following public properties:

    Id. This is an integer that uniquely identifies the role.

    Name. This is the name of the role, such as "Biologist" or "Administrator."

    Description. This is a brief description of the role.

    The MBRole class has the following public methods:

    New. This method is the default constructor.

Figure 12–42 Server MBRole class diagram

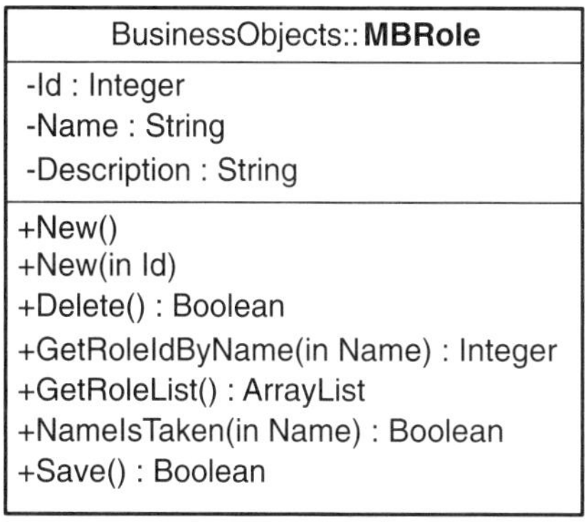

BusinessObjects::**MBRole**
-Id : Integer -Name : String -Description : String
+New() +New(in Id) +Delete() : Boolean +GetRoleIdByName(in Name) : Integer +GetRoleList() : ArrayList +NameIsTaken(in Name) : Boolean +Save() : Boolean

New (Id). This constructor queries the database to populate all of the fields based on the role's identifier, which is passed as a parameter.

Delete. This method deletes the role from the database.

GetRoleIdByName. This method looks up the role's identifier by name.

GetRoleList. This method returns an array list of MBRole objects representing every role in the database.

NameIsTaken. This method is used to check if a particular role name has already been taken. It is useful for preventing duplicates whenever you are adding a new role or changing an existing role's name.

Save. This method saves the role to the database. If the role Id is zero, the method inserts the record, retrieves the new role Id, and populates it. If the role Id is greater than zero, the method performs an update.

3. **Custom Principal class.** The Custom Principal class is used for web site security (see Figure 12–43). It implements the IPrincipal interface and is used for .NET roles authorization. Although the .NET Framework comes with a Generic Principal object, we decided to create our own Principal object and implementation of the various properties and methods. When a user logs in, he/she is issued a Forms Authentication token and a cookie with certain information, such as user identifier, username, name (first name and last name), and role. This cookie is used in the Global class's Application_Authenticate Request event. If the user is authenticated, the information is extracted from the cookie

Figure 12–43 Server Custom Principal class diagram

BusinessObjects::**CustomPrincipal**
-Id : Integer -Identity : Identity -Name : String -RoleName : String -Username : String
+New() +New(in Identity, in Id, in RoleName, in Name, in Username) +IsInRole(in RoleName)

and used to create a new Custom Principal object. This object is then placed into the HTTP Context for that web session. Our Custom Principal class contains the following public properties:

Id. This is the identifier of the user who logged in.

Identity. This object represents the identity of the user who logged in, including the user's identifier, name, username, and role name. This is read-only.

Name. This is the first name and last name of the logged-in user.

RoleName. This is the name of the logged-in user's role.

Username. This is the username of the logged-in user.

The Custom Principal class contains the following public methods:

New. This method is the default constructor.

New (Identity, Id, RoleName, Name, Username). This constructor populates all of the Custom Principal class's public properties.

IsInRole. This method checks to see if the logged-in user is in the requested role.

4. **Water Source class.** The Water Source class represents a water source that a field biologist will visit to take water samples (see Figure 12–44). It contains the following public properties:

    Id. This is an integer that uniquely identifies the water source.

    Name. This is a descriptive name of the water source (e.g., Central Park Reservoir).

    Location. This describes the location of the water source (e.g., 90th and 5th Avenue).

Figure 12–44 Server Water Source class diagram

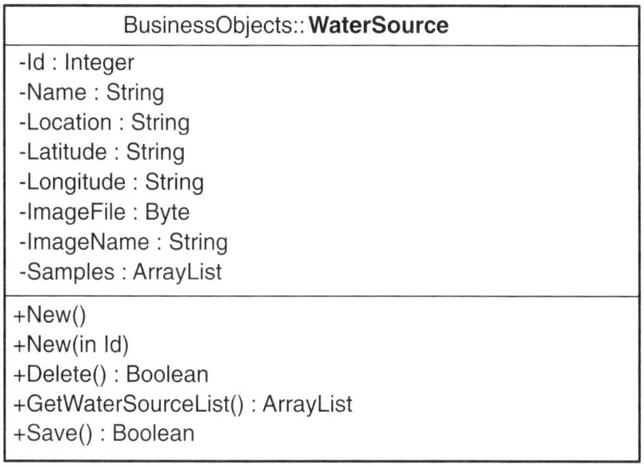

BusinessObjects:: **WaterSource**
-Id : Integer -Name : String -Location : String -Latitude : String -Longitude : String -ImageFile : Byte -ImageName : String -Samples : ArrayList
+New() +New(in Id) +Delete() : Boolean +GetWaterSourceList() : ArrayList +Save() : Boolean

Latitude and Longitude. These are coordinates that identify the exact location of the water source. These could be used in conjunction with a GPS device to pinpoint the location where the measurement was taken.

ImageFile. This is an array of bytes containing the image of the water source, uploaded by a biologist or administrator through the web site. It is only used by the Tablet PC application.

ImageName. This is the name of the image file and is used by the web site to provide a link to the image.

Samples. This is an array list of the samples that have been taken for this water source.

The Water Source class contains the following public methods:

New. This method is the default constructor.

New (Id). This constructor queries the local database to populate all of the fields based on the water source's identifier, which is passed as a parameter.

Delete. This method permanently deletes the water source from the database.

GetWaterSourceList. This method returns an array list containing all of the water sources found in the database.

Save. This method saves the water source to the database. It performs an insert if the Id is zero, or an update if it is greater than zero.

5. **Sample class.** The Sample class represents an actual water sample taken by a field biologist (see Figure 12–45). It contains the following public properties:

   Id. This positive integer uniquely identifies the sample.

   WaterSourceId. This is the unique identifier of the water source from which this sample was taken.

   SampleDate. This is the date on which the sample was taken.

   Temperature. This is the temperature of the water sample.

   pH. This is the pH level of the water sample.

   Ammonia, Nitrate, and Nitrite. These fields indicate the level of ammonia, nitrate, and nitrite present in the water sample, measured in parts per million.

   Notes. This field contains notes about the water sample.

   ImageNotes. This is an array of bytes containing the Ink markings that were made on the water source image to indicate where the sample was taken. These are entered via the Tablet PC application.

Figure 12–45 Server Sample class diagram

BusinessObjects::**Sample**
-Id : Integer -WaterSourceId : Integer -SampleDate : Date -Temperature : Decimal -pH : Decimal -Ammonia : Decimal -Nitrate : Decimal -Nitrite : Decimal -Notes : String -ImageNotes : Byte
+New() +New(in Id) +Delete() : Boolean +GetSampleListByLocation() : ArrayList +GetSampleReport() : ArrayList +Save() : Boolean

The Sample class contains the following public methods:

New. This method is the default constructor.

New (Id). This constructor queries the database to populate all of the fields based on the sample's identifier, which is passed as a parameter.

Delete. This method permanently deletes the sample from the database.

GetSampleListByLocation. This method returns an array list of all samples at a given water source.

GetSampleReport. This method returns an array list of all samples that meet the report parameters.

Save. This method saves the sample to the database. It performs an insert if the Id is zero, or an update if it is greater than zero.

12.5.4 Data Access Layer

The server application uses Microsoft SQL Server 2000. All communication with the database is carried out through objects in the MobileBiologist.DataAccessObjects namespace (see Figure 12–46). This section describes those objects and the actual database schema.

Figure 12–46 Server Data Access Objects

DataAccessObjects::**DataConnHelper**
+getStrConn() : String

DataAccessObjects::**SQLHelper**

DataAccessObjects::**SQLParameterCache**

1. **Data access objects.** The DataConnHelper class is used to look up the database connection string, which is then passed by the business objects to the SQLHelper class. The SQLHelper and SQLParameterCache classes are part of the Data Access Application Block for .NET and were provided by Microsoft for general use.
2. **Database design.** The server database was implemented using Microsoft SQL Server 2000. Figure 12–47 illustrates the database tables.

Figure 12–47 Server database tables

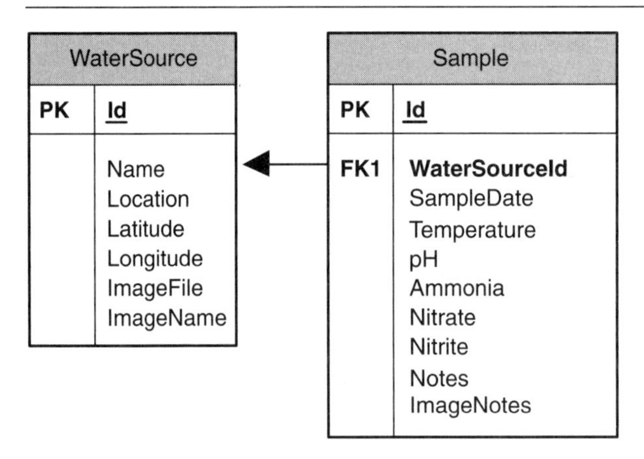

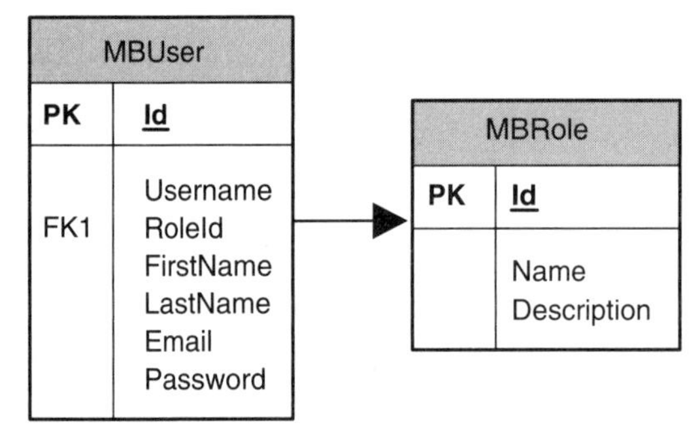

The Water Source table contains the following fields:

Field Name	Data Type	Primary Key
Id	int	Yes
Name	varchar(255)	
Location	varchar(255)	
Latitude	varchar(50)	
Longitude	varchar(50)	
ImageFile	image	
ImageName	varchar(255)	

The Sample table contains the following fields:

Field Name	Data Type	Primary Key
Id	int	Yes
WaterSourceId	int	Foreign key
SampleDate	smalldatetime	
Temperature	float	
pH	float	
Ammonia	float	
Nitrate	float	
Nitrite	float	
Notes	varchar(255)	
ImageNotes	image	

The MBUser table contains the following fields:

Field Name	Data Type	Primary Key
Id	int	Yes
Username	varchar(50)	
RoleId	int	Foreign key
FirstName	varchar(50)	
LastName	varchar(50)	
Email	varchar(255)	
Password	varchar(50)	

The MBRole table contains the following fields:

Field Name	Data Type	Primary Key
Id	int	Yes
Name	varchar(50)	
Description	varchar(255)	

The database contains the following stored procedures:

Procedure Name	Parameters	Return Value
sp_AddMBRole	@Name varchar(50), @Description varchar(255)	Id
sp_AddMBUser	@Username varchar(50), @FirstName varchar(50), @LastName varchar(50), @Email varchar(255), @Password varchar(50), @RoleId int	Id
sp_AddSample	@WaterSourceId int, @SampleDate smalldatetime, @Temperature float, @pH float, @Ammonia float, @Nitrate float, @Nitrite float, @Notes varchar(255), @ImageNotes image=null	Id
sp_AddWaterSource	@Name varchar(255), @Location varchar(255), @Latitude varchar(50), @Longitude varchar(50), @ImageName varchar(255)=null, @ImageFile image=null	Id
sp_UpdateMBRole	@Id int, @Name varchar(50), @Description varchar(255)	

Procedure Name	Parameters	Return Value
sp_UpdateMBUser	@Id int, @Username varchar(50), @FirstName varchar(50), @LastName varchar(50), @Email varchar(255), @Password varchar(50), @RoleId int	
sp_UpdateSample	@Id int, @WaterSourceId int, @SampleDate smalldatetime, @Temperature float, @pH float, @Ammonia float, @Nitrate float, @Nitrite float, @Notes varchar(255), @ImageNotes image=null	
sp_UpdateWaterSource	@Id int, @Name varchar(255), @Location varchar(255), @Latitude varchar(50), @Longitude varchar(50), @ImageFile image=null, @ImageName varchar(255) =null	

12.6 DISCUSSION

The Mobile Biologist application provides an example of a fat client application that uses a store-and-forward mechanism. It can be applied to a variety of situations in which a person has to gather information while periodically disconnected from the server application, but also needs to continue working and synchronize that information with the server application later. This type of scenario is very common in many areas today. Some examples are described below.

Healthcare Industry. Mobile healthcare workers may use such applications to gather patient information while on house calls, where network access is far from guaranteed. Later, when the workers return to the office, they may upload the data to a back-end server.

Government. Government field employees may use an application like this to perform disease tracking when there is a virus outbreak, such as West Nile virus or SARS. They may gather epidemiological information from remote locations and enter data on their mobile devices when network access is not guaranteed. Later, when they return to the office, they may upload the data to a back-end server for more detailed analysis.

12.7 EXTENSIONS

This application could be extended to provide additional functionality through the use of special hardware attachments and software, such as those described below.

Biometrics. For added security, you could use biometric identification on the mobile device. For example, many Pocket PCs have a built-in thumbprint identification mechanism while a separate biometric identification card can be purchased for the Tablet PC. In the application we developed, this is probably not necessary. However, biometric identification might be extremely useful for applications that store more sensitive data, such as patient medical information.

GPS. One nice enhancement to this application would be the addition of GPS hardware and software. For example, you could buy a sleeve that fits over a Pocket PC that enables you to use GPS. You can also purchase wireless GPS receivers that communicate with a Pocket PC or a Tablet PC using Bluetooth. GPS could be used to help the field biologists navigate to the sites they are going to visit or automatically pinpoint the exact location where a water sample was taken.

Camera. Another possible enhancement to this application would be the use of a camera on the mobile device. For example, HP makes a mobile camera that attaches to a Pocket PC using the SD slot, and ViewSonic makes a Tablet PC with a camera built in. Camera functionality would allow biologists to actually take a picture of the water source they are sampling. It also might be useful for recording weather conditions, the water level, contamination, and the condition of the water.

Handwritten Notes. The Tablet PC application could be extended to provide additional functionality specific to the Tablet PC. For example, although we utilized the InkEdit and InkPicture controls, we did not use the full Ink capabilities of the Tablet PC. For example, on the Tablet PC, handwriting can be captured as digital Ink, which is not converted to regular typefaced text. This would allow the biologists to write and save notes in their own handwriting.

12.8 SUMMARY

This chapter described the development of a fat client mobile application that can be used by biologists to enter water sample data while they are in the field, using either a Pocket PC or a Tablet PC.

We initially described the application's requirements. For example, biologists can download data while they are connected to the network, add information while they are disconnected, and upload data when they return to the office.

We then described the architecture of the overall solution along with the detailed design of the client and server applications. The Pocket PC client was developed using Microsoft SQL Server CE, VB.NET, and the .NET Compact Framework. The Tablet PC client was developed using Microsoft SQL Server 2000, VB.NET, and the full .NET Framework. The server was developed using Microsoft SQL Server 2000, ASP.NET, VB.NET, and the full .NET Framework. Data was synchronized through the use of server business objects that were exposed as web services using SOAP.

Finally, we discussed other areas where fat client applications can be used. We also considered some of the extensions that might be added to this application, including biometric authentication, GPS location tracking, digital photography, and storage of handwritten notes.

CHAPTER 13

Mobile Zoo Case Study

Our greatest glory is not in never falling,
but in rising every time we fall.
—Confucius

In this case study, we discuss an application called the "Mobile Zoo." This application allows zoo visitors to use Pocket PCs to view web pages that contain animal and zoo information while they are touring the zoo. Visitors can also sign up to receive zoo-related information or donate money to the zoo. All of these activities are possible even when visitors are disconnected from the zoo's back-end web server.

The unusual aspect of this case study is that the application that gathers data on the Pocket PC is not a traditional fat client Windows Forms application. Instead, the entire application consists of a set of web pages hosted on the Pocket PC mobile client, using a program we developed called the Pocket Web Host (PWH).

PWH is a "mini" web server that enables us to service web pages on a Pocket PC even when it is disconnected from a back-end web server. The internals of PWH will not be described in this chapter. Readers who are interested in specific details of this case study and PWH may download the full code listing from the companion web site.

13.1 USE CASES

Suppose that you are responsible for the visitor education center of a zoo and you have an existing Internet web site that provides information about the animals exhibited in the zoo. The web site contains a map of the zoo along with descriptions of the animals encountered on recommended walking routes throughout the zoo. The web site also allows visitors to register with the zoo in order to receive newsletters and other assorted zoo-related information.

You would like to provide zoo visitors with a Pocket PC that contains the same information as the original web site and also allows the gathering of the same visitor information. Visitors can also use this Pocket PC to take self-guided tours of the zoo.

Visitors will rent the Pocket PC at the zoo kiosk entrance for a nominal charge, leave a credit card number as a deposit, and walk around the zoo with the Pocket PC, which will give them information about the animals in the zoo. Visitors can also fill out a form that allows them to register for various zoo-related services.

When a visitor is ready to leave, he/she will return the Pocket PC to the zoo kiosk. A zoo employee will connect the Pocket PC to the network using a wired connection or a Pocket PC cradle. He/she will then upload any data entered on the Pocket PC into the server database and delete the uploaded data from the Pocket PC.

In Chapter 11, we developed a similar application for the Mobile Museum Guide. In that example, however, we assumed that the Pocket PCs would be wirelessly networked into the museum's network and that a thin client model (a web or WAP browser) would be used.

However, while it might be relatively simple to run a complete wireless network in a museum, it might be impossible to do the same in a zoo because of its size. For example, it might be prohibitively expensive to implement a wireless network that provides constant coverage throughout the entire zoo.

The problem is that we want the Pocket PC to display web pages even though it will not be connected to a wireless network. Ordinarily, to support a mobile device that is not connected to a network, you would need to develop a fat client Windows Forms application. However, the existing web site has a large number of images and text and it would be extremely difficult to rewrite them using Windows Forms. As a result we decided to develop a mechanism to host web pages on the Pocket PC, which we called the Pocket Web Host (PWH). For the purpose of this case study, you can simply treat this as a mini web-server that resides on the Pocket PC.

13.1.1 Use Case Actors

The current zoo web site supports users in two different roles. When developing use cases for the Mobile Zoo web site, we considered each of these roles and how they might be portrayed as actors in use cases (see Figure 13–1). There are two types of actor as described below:

1. **Prospects.** Prospects are visitors to the web site who are able to view self-guided walking tours and information about the animals in the zoo. They may also submit information about themselves and even make a donation to the zoo. We refer to these users as

Figure 13–1 Use case actors

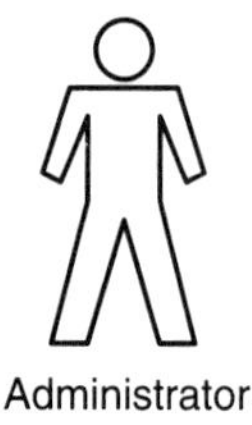

prospects since each user is a prospective zoo member or benefactor. One of the primary goals of this application is to provide prospects the opportunity to enter their personal information, credit card information, and a donation amount.

2. **Administrators.** Administrators are able to add, edit, and delete other users and roles. They can also add, edit, and delete the data submitted by prospects.

The following sections describe the use cases involving each of the two actors.

13.1.2 Existing Web Site Use Cases

The existing web site currently supports a number of use cases. Some of these may be performed by members of the public who visit the web site from their home computers. Others may be performed by administrators who work at the zoo. Prospects may perform the following tasks on the web site (see Figure 13–2):

View self-guided walks. Web site visitors may choose from several self-guided walking tours of the zoo. Each self-guided walk contains a map of the route, as well as pictures and descriptions of the animals encountered along the way.

View help. All users may view a help page, which provides helpful information about the zoo and web site.

Submit personal information. All users may submit their personal information to sign up for various zoo services. If users want to make a donation, they may also submit their credit card information and gift amount.

Figure 13–2 Prospect use cases

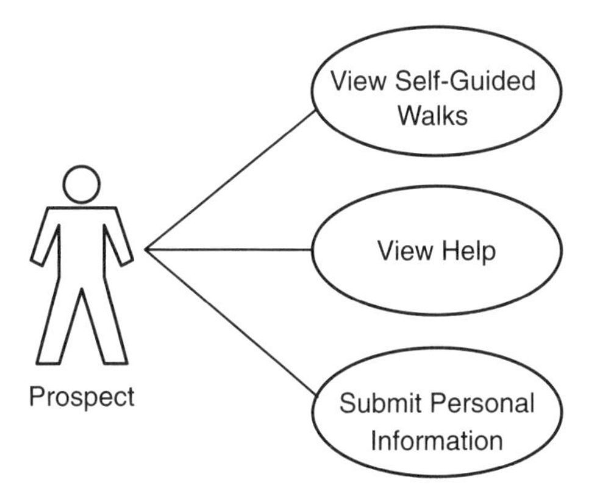

Administrators may perform a variety of additional tasks on the web site (see Figure 13–3):

Log in. Administrators must enter a username and password to log in to the administrative area of the web site.

Add/edit/delete users. Administrators may add, edit, and delete web site user accounts.

Add/edit/delete roles. Administrators may add, edit, and delete web site user roles.

Add/edit/delete prospects. Administrators may add, edit, and delete the information entered by prospects. For example, administrators may visit the web site each day to see how many new prospective members have expressed an interest in the zoo or have made a donation.

13.1.3 Mobile Web Site Use Cases

When you mobilize an existing application, you must decide which use cases to support in the mobile client application. In this case, we decided to mobilize all of the prospect use cases, as described below:

View self-guided walks. We thought it was important to support this use case since a primary goal of the mobile application is to allow users to take self-guided tours of the zoo using the Pocket PC.

View help. Although not strictly necessary, we decided to keep the help function since this is a very simple use case and should be easy to support in a mobile application. In addition, this may allow us to provide help specifically tailored to the mobile web site.

Figure 13–3 Administrator use cases

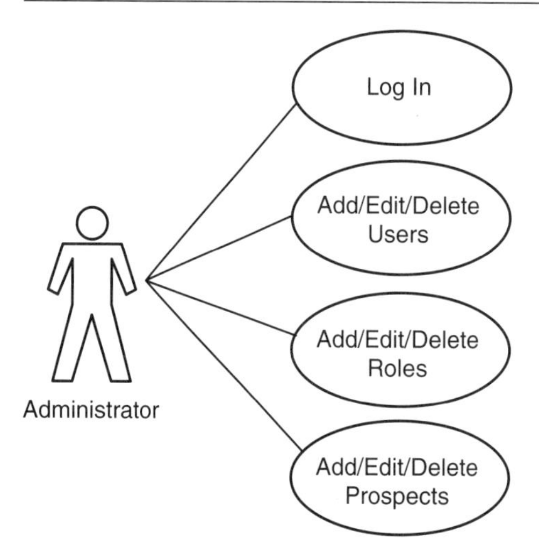

Submit personal information. We decided to keep this use case since the other important goal of creating the Mobile Zoo web site is to provide a useful service to visitors that will hopefully entice them to make a donation to the zoo. Visitors are more likely to donate if they can make the donation on their mobile device before leaving the zoo rather than waiting until they get home.

We decided not to mobilize the administrative use cases, since these functions will typically be performed by an administrator who is working in an office using a desktop or laptop computer.

13.2 ARCHITECTURE

In the following section, we describe the architecture of the Mobile Zoo application. Please note that we have implemented a simple set of web sites for demonstration purposes only. In enterprise-wide web sites, multiple servers, security, multi-tiered web architectures, and other considerations must be taken into account.

13.2.1 Overall Architecture

Figure 13–4 illustrates the overall architecture of the application. The Pocket PC application contains the following items:

- Web Pages
- Business Objects
- Data Access Objects
- Database
- PWH

The Web Server contains the following items:

- HTML Content
- ASP.NET Web Forms
- XML Web Service
- Business Objects
- Data Access Objects
- Database

Each item, except for PWH, is discussed in detail throughout the rest of this chapter. The design of PWH is discussed in Appendix B.

Figure 13–4 Mobile Zoo architecture

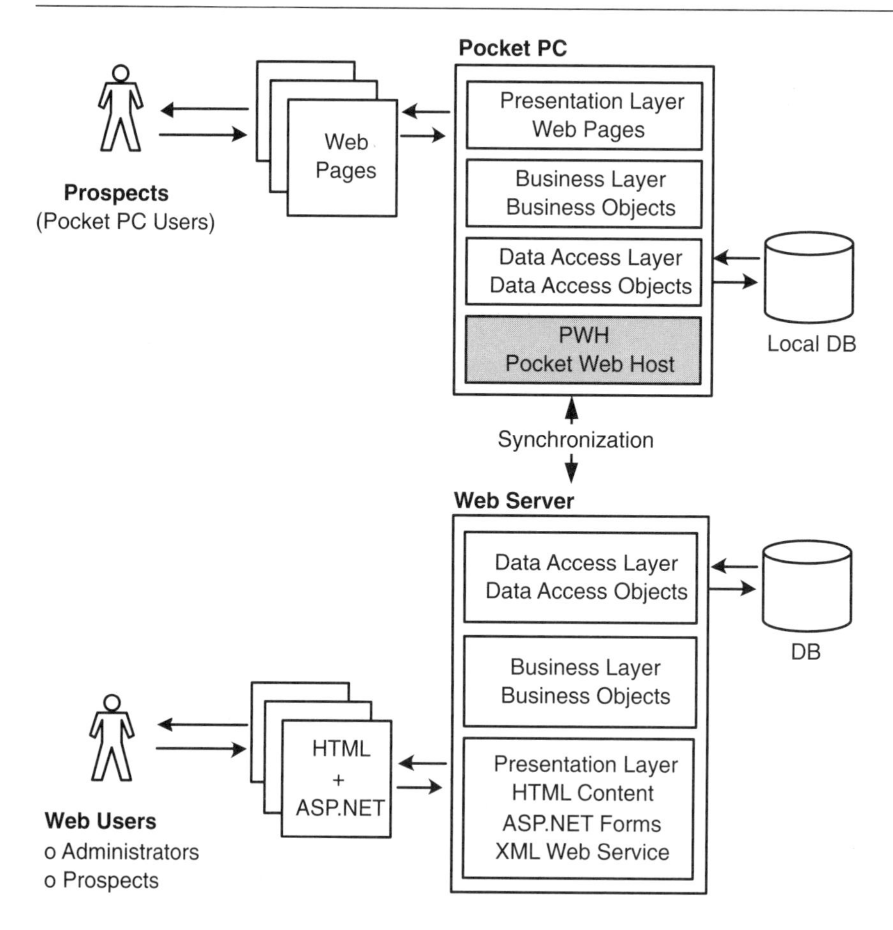

13.3 CLIENT DETAILED DESIGN

This section describes the detailed design of the Pocket PC application, including its class diagrams and object models.

All objects in the Pocket PC client application are in the namespace "MobileZoo.PPC." As shown in Figure 13–5, we have created several different sub-namespaces to hold the various objects.

13.3.1 Presentation Layer

The Presentation Layer consists primarily of HTML pages that display images of the zoo and animals (see Figure 13–6). It also contains two .PWH web pages with their associated code-behind files, which were developed using C#. These pages are designed to work with PWH and the .NET

Figure 13–5 Namespaces in the Mobile Zoo Pocket PC application

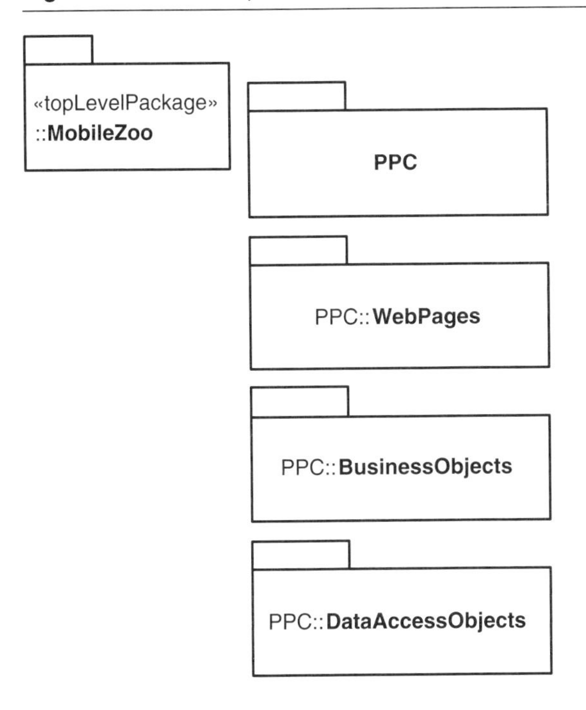

Figure 13–6 Web pages in the Pocket PC application

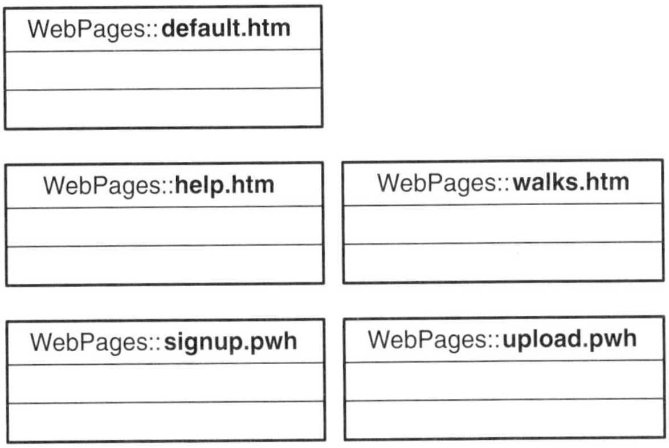

WebPages::**default.htm**

WebPages::**help.htm**

WebPages::**walks.htm**

WebPages::**signup.pwh**

WebPages::**upload.pwh**

Compact Framework. We created code-behinds as classes in the MobileZoo.PPC.WebPages namespace. The rest of this section describes each page in more detail.

1. **Default page.** The default page, default.htm, welcomes visitors to the web site and explains the application (see Figure 13–7).

Figure 13–7 Default page

2. **Help page.** The help page, help.htm, provides helpful information about the zoo and the web site (see Figure 13–8).

Figure 13–8 Help page

3. **Walks page.** The walks page, walks.htm, allows a user to select a self-guided walking tour of the zoo (see Figure 13–9). For demonstration purposes, we provide one walk, which is a clockwise tour of the zoo. The clockwise walk page shows a map of the walk-

Figure 13–9 Walks page

ing route (see Figure 13–10). This walk contains seven additional HTML pages, all of which show pictures and information on each of the different animals seen en route (see Figure 13–11 for an example).

Figure 13–10 Clockwise walk page

Figure 13–11 Iguana information page

4. **Signup page.** The signup page, signup.pwh, uses the code-behind file Signup_CB.cs. This page allows a user to submit all of his/her personal and credit card information to the local database (see Figure 13–12).

Figure 13–12 Signup page

5. **Upload page.** The upload page, upload.pwh, uses the code-behind file Upload_CB.cs. This page uploads all new data to the server, deletes the data from the local database, and notifies the user that the upload is complete. This page is used by the zoo administrator

when a visitor returns the Pocket PC to the zoo kiosk. This page will only allow an upload if the user is connected to the network (see Figure 13–13).

Figure 13–13 Upload page

13.3.2 Business Logic Layer

This section describes the objects in the MobileZoo.PPC.BusinessObjects namespace.

1. **MZ_Util class.** The MZ_Util class contains a variety of helper functions that are used by the Pocket PC web page code-behind files (see Figure 13–14). This class contains no public properties and does not have a corresponding class in the server application. It contains the following public methods:

    New. This method is the default constructor.

    AddProspect. This method adds a prospect to the local database.

    AddProspectsToServer. This method uploads all new prospect information to the server database using the web service.

    CheckConnection. This method determines whether or not the Pocket PC has an active network connection by calling a method on the web service.

    CheckForChanges. This method determines whether or not the user has added prospect information to the local database.

    CheckInitialized. This method determines whether or not the local database has been created.

    CleanDatabase. This method deletes all of the existing prospect information from the local database after it has been uploaded to the server.

    GetDBName. This method returns the name of the database.

    GetProspects. This method returns a dataset containing all of the prospects in the local database.

Figure 13–14 Utility Class Diagram

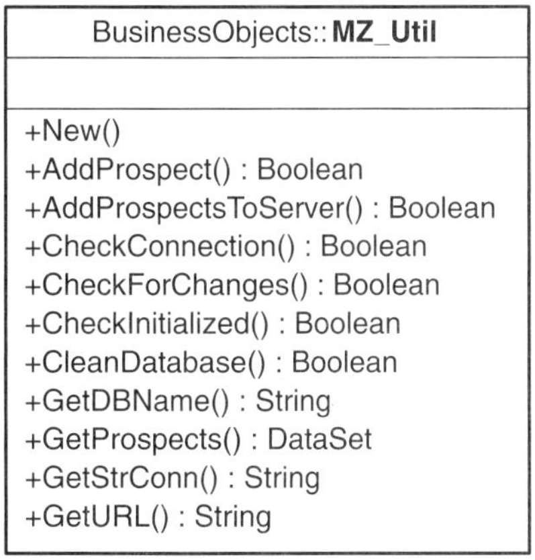

BusinessObjects::**MZ_Util**
+New() +AddProspect() : Boolean +AddProspectsToServer() : Boolean +CheckConnection() : Boolean +CheckForChanges() : Boolean +CheckInitialized() : Boolean +CleanDatabase() : Boolean +GetDBName() : String +GetProspects() : DataSet +GetStrConn() : String +GetURL() : String

GetStrConn. This method returns the database connection string.

GetURL. This method returns the URL of the web service. This is found in the file MZ_config.xml, located in the Mobile Zoo directory under the PWH deployment area.

2. **Server Proxy class.** The Server Proxy class communicates with the web service on the remote server (see Figure 13–15). This class is auto-generated by Microsoft Visual Studio .NET when a web reference is added. This class contains a default URL that is set when the class is created. This URL can be overridden after an instance of the class is created. The URL is stored in the MZ_config.xml file, which gets deployed with the application. This makes it possible to change the URL without rebuilding the application. The Server Proxy class will only need to change if the web service changes. In that case, the web reference should be updated and the class will be regenerated. This class has no public properties. It has the following public methods:

    New. This constructor sets the default URL of the web service. This property is reset by the application after an instance of the class is created.

    AddProspects. This method takes an XML string containing all of the prospect information added locally and sends it to the server so that it can be added to the remote database.

    EchoString. This method sends a string to the remote server, which returns a date. It is used to test the web service and see if a connection can be established with the server.

Ideally, you should set up SSL on your web server so that all calls to the web service can be made using HTTPS and all the data sent will be encrypted. However, for simplicity, we did not set up SSL on the web server for this case study.

Figure 13–15 Server Proxy class diagram

Server:: **MZ_WS**
+New() +AddProspects(in prospectXML) : Boolean +EchoString(in strval)

13.3.3 Data Access Layer

All communication with the local database is carried out through an object in the MobileZoo.PPC.DataAccessObjects namespace. This section describes that object and the actual database schema. The database file mz_db.sdf is created by the program MZ_DB_Setup and deployed with the Mobile Zoo application.

1. **SQLCEHelper class.** This class provides a set of functions for manipulating the local database and performing queries (see Figure 13–16). It is a scaled-down version of the SQLHelper class, which is used by the server application. The SQLCEHelper class has no public properties. It has the following public methods:

   New. This method is the default constructor.

   CreateDB. This method creates a database with the connection string passed as a parameter. This connection string includes the database name and password.

   DBExists. This method checks to see if the specified database exists.

   DeleteDB. This method deletes the specified database.

   ExecuteCommand. This method connects to the specified database and executes the given SQL statement.

   ExecuteDataset. This method executes the given SQL statement and returns a dataset.

   ExecuteScalar. This method executes the given SQL statement and returns the scalar value returned by the SQL call.

Figure 13–16 SQLCEHelper class diagram

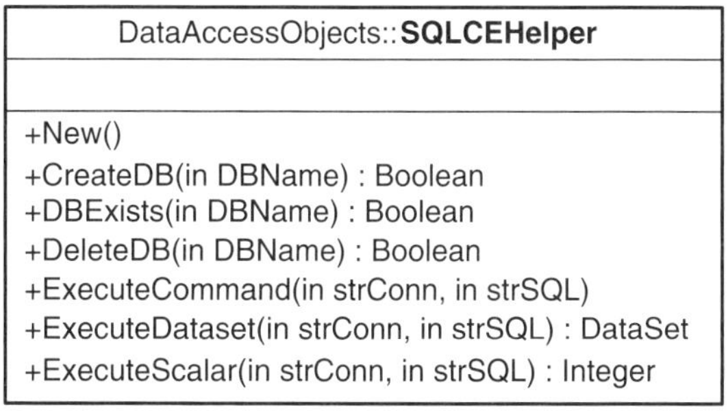

DataAccessObjects::**SQLCEHelper**
+New() +CreateDB(in DBName) : Boolean +DBExists(in DBName) : Boolean +DeleteDB(in DBName) : Boolean +ExecuteCommand(in strConn, in strSQL) +ExecuteDataset(in strConn, in strSQL) : DataSet +ExecuteScalar(in strConn, in strSQL) : Integer

2. **Database schema.** The local database contains one table, the Prospect table. The Prospect table contains the following fields:

Field Name	Data Type	Primary Key
Id	int	Yes
FirstName	nvarchar(50)	
MiddleInitial	nvarchar(50)	
LastName	nvarchar(50)	
Address	nvarchar(50)	
Apartment	nvarchar(50)	
City	nvarchar(50)	
State	nvarchar(50)	
Zip	nvarchar(50)	
Country	nvarchar(50)	
DaytimePhone	nvarchar(50)	
EmailAddress	nvarchar(50)	
GiftAmount	nvarchar(50)	
DataComplete	bit	
AccountNumber	nvarchar(50)	
CreditCard	nvarchar(50)	
ExpDateMonth	int	
ExpDateYear	int	
LastUpdateDate	datetime	

13.4 SERVER DETAILED DESIGN

All server objects are in the namespace "Mobile Zoo." As shown in Figure 13–17 we have created several different sub-namespaces to hold the various objects.

13.4.1 XML Web Service

The web service provides a mechanism for the Pocket PC to communicate with the server. There is only one class in this namespace, MZ_WS (see Figure 13–18). The name MZ_WS stands for "Mobile Zoo Web Service." It has no public properties. It includes the following public methods:

New. This method is the default constructor.

EchoString. This is a simple method that returns a date; it is used to test connectivity to the web service.

AddProspects. This method takes an XML string that contains the new prospects from the Pocket PC. It iterates through the XML document and creates new Prospect business objects, populating them with the XML data. It then saves each object to the database.

Figure 13–17 Namespaces in the Mobile Zoo server application

Figure 13–18 MZ_WS class diagram

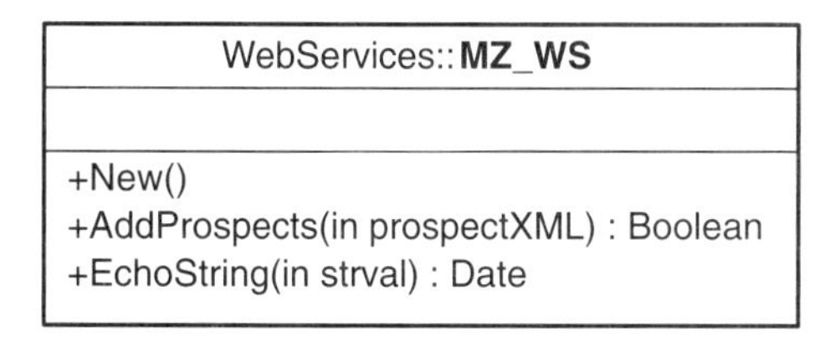

13.4.2 Presentation Layer

The Presentation Layer consists of a set of HTML and ASP.NET web pages and user controls, as described below.

1. **Utility web pages and controls.** Figure 13–19 illustrates the utility web pages and controls, including:

    Global. This utility web page replaces the traditional Global.asa file in ASP applications. The Global class provides event handlers such as Session_Start, Application_Start, Application_AuthenticateRequest, etc.

Figure 13–19 Utility web pages and controls

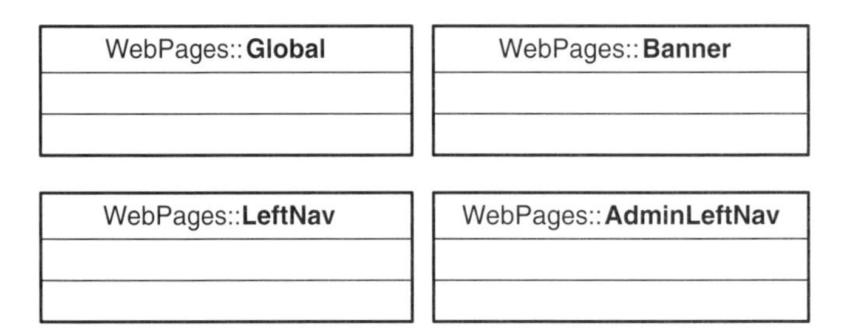

The Application_AuthenticateRequest method creates a new Custom Principal object and adds it to the current HTTP Context.

Banner. This web user control displays a banner at the top of the page.

LeftNav. This web user control displays a left-hand navigation bar. It checks the role of the logged-in user to decide which links to display.

AdminLeftNav. This web user control displays a left-hand navigation bar for the secure administrative area of the web site.

2. **Default Web pages.** Figure 13–20 illustrates the default web pages that are available to all users. This web site is protected by Forms Authentication. However, the default pages can be accessed by anyone. This is specified in the Web.config file.

```
<authentication mode="Forms">
<forms loginUrl="/MZ_WebSite/login.aspx"
name="sqlAuthCookie" timeout="60" path="/"></forms>
</authentication>

<authorization>
<allow users="*" /> <!– Allow all users-->
</authorization>
```

Default Page. This is the home page of the website.

Help. This page shows help for the web site.

Signup. This page allows visitors to sign up to receive marketing materials or make a donation.

Walks. This page shows a list of the different walking tours that are available. Walking tours contain pages describing several different animals.

Figure 13–20 Default web pages

WebPages:: **defaultPage**	WebPages:: **help**

WebPages::**walks**	WebPages::**signup**

3. **Administrative web pages.** There are several pages that are only accessible to administrators (see Figure 13–21). These pages are used to add, update, and delete users, roles and prospects. To secure these pages, we placed them in a subdirectory called "admin." We also placed a Web.config file in that directory, which indicates that the pages are only accessible to users in the role "ADMIN."

```
<authorization>
<deny users="?"/>
<allow roles="ADMIN"/>
<deny roles="DEFAULT"/>
</authorization>
```

The following describes the web pages that are only accessible to administrators:

Login. Although this page is accessible to all users, it is only used by administrators, since they are the only ones required to log in.

The login page allows users to enter a username and password. If the information entered is valid, the page issues the Forms Authentication token. It also sets a cookie containing user information. This cookie is used by the Global class's Application_AuthenticateRequest method to create the Custom Principal object. The Custom Principal object is used by the Forms Authentication filter to perform role-based security.

The login page is also used to log users out of the web site by calling the FormsAuthentication.SignOut method, deleting the user information cookie, and removing the user from the HTTP Context.

AdminHome. This is the home page of the administration section of the site.

Users. This page allows the administrator to view a list of users.

Roles. This page allows the administrator to view a list of roles.

Figure 13–21 Administrator web pages

WebPages:: **login**

WebPages:: **adminHome**

WebPages:: **users**

WebPages:: **roles**

WebPages:: **prospects**

WebPages:: **edit_user**

WebPages:: **edit_role**

WebPages:: **edit_prospect**

Prospects. This page allows the administrator to view a list of prospects who have submitted information through the server or mobile web site.

Edit User. This page allows the administrator to add or edit a user's information, including what role the user belongs to.

Edit Role. This page allows the administrator to add or edit a role's information, such as its name and description.

Edit Prospect. This page allows the administrator to add or edit a prospect's information.

13.4.3 Business Logic Layer

This section describes all the objects in the MobileZoo.BusinessObjects namespace.

1. **MZUser class.** This class represents a user of the Mobile Zoo application (see Figure 13–22). It has the following public properties:

   Id. This is an integer that uniquely identifies the user.

   FirstName. This is the user's first name.

   LastName. This is the user's last name.

   Name. This is a combination of the user's first and last names. It is read-only and used for display purposes.

Figure 13–22 MZUser class diagram

DataAccessObjects::**MZUser**
-Id : Integer -FirstName : String -LastName : String -Name : String -RoleId : Integer -RoleName : String -Email : String -Username : String -Password : String
+New() +New(in Id) +Delete() : Boolean +GetUserList() : ArrayList +Login(in Username, in Password) : Boolean +NameIsTaken(in Name) : Boolean +Save() : Boolean

RoleId. This is the identifier of the role the user belongs to.

RoleName. This is the name of the role the user belongs to.

Email. This is the user's email address.

Username. This is the username used to log in to the web site.

Password. This is the user's password. For purposes of this sample application, it is not encrypted, although it could be encrypted using the Forms Authentication object's HashPasswordForStoringInConfigFile method.

The MZUser class has the following public methods:

New. This method is the default constructor.

New (Id as integer). This constructor queries the database to populate all of the fields based on the user's identifier, which is passed as a parameter.

Delete. This method deletes the user from the database.

GetUserList. This method returns an array list of MZUser objects representing every user in the database.

Login. This method verifies the username and password against the value in the database.

NameIsTaken. This method is used to check if a particular username has already been taken. It is useful for preventing duplicates whenever you are adding a new user or changing an existing user's username.

Save. This method saves the user's data to the database. If the user Id is zero, the method inserts the record, retrieves the new user Id, and populates it. If the user Id is greater than zero, the method performs an update.

2. **MZRole class.** The MZRole class represents a security role for the Mobile Zoo web application (see Figure 13–23). It has the following public properties:

   Id. This is an integer that uniquely identifies the role.

   Name. This is the name of the role, such as "ADMIN".

   Description. This is a brief description of the role.

   The MZRole class has the following public methods:

   New. This method is the default constructor.

   New (Id). This constructor queries the database to populate all of the fields based on the role's identifier, which is passed as a parameter.

   Delete. This method deletes the role from the database.

   GetRoleIdByName. This method looks up the identifier for a role, given its name.

   GetRoleList. This method returns an array list of MZRole objects representing every role in the database.

   NameIsTaken. This method is used to check if a particular role name has already been taken. It is useful for preventing duplicates whenever you are adding a new role or changing an existing role's name.

   Save. This method saves the role to the database. If the role Id is zero, the method inserts the record, retrieves the new role Id, and populates it. If the role Id is greater than zero, the method performs an update.

Figure 13–23 MZRole class diagram

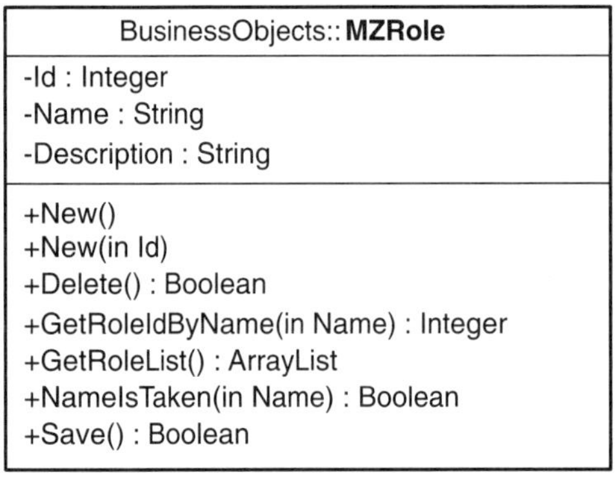

BusinessObjects:: **MZRole**
-Id : Integer -Name : String -Description : String
+New() +New(in Id) +Delete() : Boolean +GetRoleIdByName(in Name) : Integer +GetRoleList() : ArrayList +NameIsTaken(in Name) : Boolean +Save() : Boolean

3. **Custom Principal class.** The Custom Principal class is used for web site security (see Figure 13–24). It implements the IPrincipal interface and is used for .NET roles authorization. Although the .NET Framework comes with a Generic Principal object, we decided to create our own Principal object and implementation of the various properties and methods. When a user logs in, he/she is issued a Forms Authentication token and a cookie with certain information, such as his/her user identifier, username, name (first name and last name) and role. This cookie is used in the Global class's Application_AuthenticateRequest event. If the user is authenticated, the information is extracted from the cookie and used to create a new Custom Principal object. This object is then placed into the HTTP Context for that web session. Our Custom Principal class contains the following public properties:

   Id. This is the identifier of the user who logged in.

   Identity. This object represents the identity of the user who logged in. It includes the user's identifier, name, username, and role name. This is read-only.

   Name. This is the first name and last name of the logged-in user.

   RoleName. This is the name of the role of the logged-in user.

   Username. This is the username of the logged-in user.

   The Custom Principal class contains the following public methods:

   New. This method is the default constructor.

   New (Identity, Id, RoleName, Name, Username). This constructor populates all of the Custom Principal class's public properties.

   IsInRole. This method checks to see if the logged-in user is in the requested role.

Figure 13–24 Custom Principal class diagram

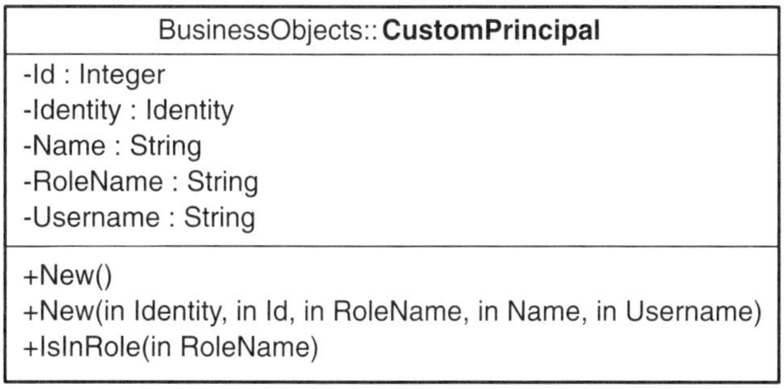

BusinessObjects::**CustomPrincipal**
-Id : Integer -Identity : Identity -Name : String -RoleName : String -Username : String
+New() +New(in Identity, in Id, in RoleName, in Name, in Username) +IsInRole(in RoleName)

4. **Prospect class.** The Prospect class represents a visitor to the zoo web site who is a prospective member or benefactor (see Figure 13–25). It contains the following public properties:

    Id. This is an integer that uniquely identifies the prospect.

    FirstName. This is the prospect's first name.

    LastName. This is the prospect's last name.

    MiddleInitial. This is the prospect's middle initial.

    Address. This is the prospect's address.

    Apartment. This is the prospect's apartment number.

    City. This is the prospect's city.

    State. This is the prospect's state.

    Zip. This is the prospect's ZIP code.

    Country. This is the prospect's country of residence.

    DaytimePhone. This is the prospect's phone number.

    EmailAddress. This is the prospect's email address.

    GiftAmount. This is the amount of the donation the prospect would like to give to the zoo.

    DataComplete. This is a flag indicating whether or not the prospect's data is complete.

    AccountNumber. This is the prospect's credit card account number.

    CreditCard. This is the prospect's credit card type, such as Visa or MasterCard.

    ExpDateMonth. This is the month the prospect's credit card expires.

    ExpDateYear. This is the year the prospect's credit card expires.

    Last Update Date. This is the date the record was last updated.

    The Prospect class contains the following public methods:

    New. This method is the default constructor.

    New (Id). This constructor queries the local database to populate all of the fields based on the prospect's identifier, which is passed as a parameter.

    Delete. This method permanently deletes the prospect from the database.

    GetProspectList. This method returns an array list containing all of the prospects found in the database.

    Save. This method saves the prospect to the database. It performs an insert if the Id is zero, or an update if it is greater than zero.

Figure 13–25 Prospect class diagram

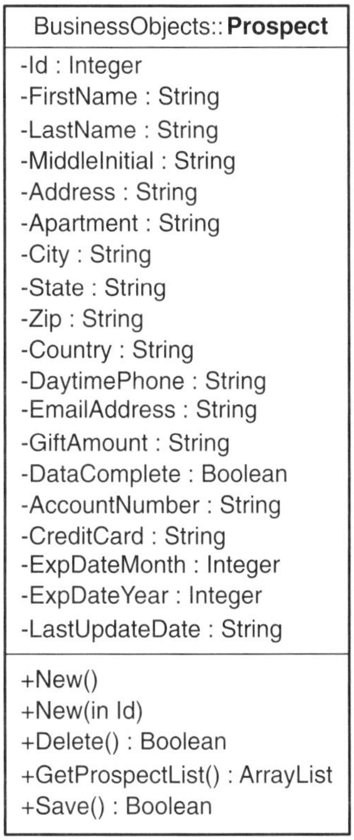

BusinessObjects::**Prospect**
-Id : Integer -FirstName : String -LastName : String -MiddleInitial : String -Address : String -Apartment : String -City : String -State : String -Zip : String -Country : String -DaytimePhone : String -EmailAddress : String -GiftAmount : String -DataComplete : Boolean -AccountNumber : String -CreditCard : String -ExpDateMonth : Integer -ExpDateYear : Integer -LastUpdateDate : String
+New() +New(in Id) +Delete() : Boolean +GetProspectList() : ArrayList +Save() : Boolean

13.4.4 Data Access Layer

The server application uses Microsoft SQL Server 2000. All communication with the database is carried out through objects in the MobileZoo.DataAccessObjects namespace (see Figure 13–26). This section describes these objects and the actual database schema.

1. **Data access objects.** The DataConnHelper class is used to look up the database connection string, which is then passed by the business objects to the SQLHelper class. The SQLHelper and SQLParameterCache classes are part of the Data Access Application Block for .NET and were provided by Microsoft for general use.
2. **Database design.** The server database was implemented using Microsoft SQL Server 2000. Figure 13–27 illustrates the database tables.

Figure 13–26 Mobile Zoo data access objects

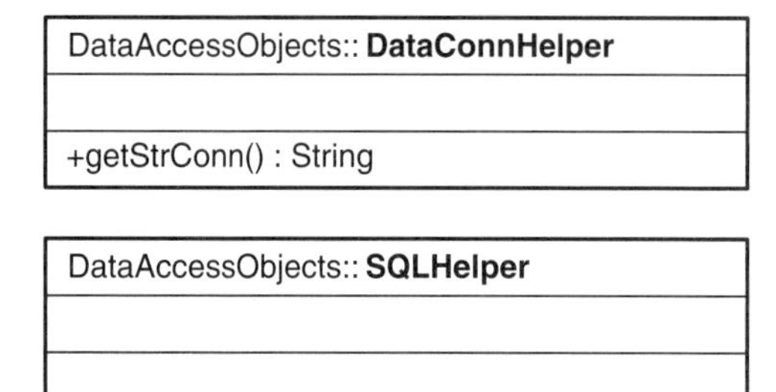

DataAccessObjects:: **SQLParameterCache**

Figure 13–27 Database tables

MZUser	
PK	**Id**
	Username
FK1	RoleId
	FirstName
	LastName
	Email
	Password

MZRole	
PK	**Id**
	Name
	Description

Prospect	
PK	**Id**
	FirstName
	MiddleInitial
	LastName
	Address
	Apartment
	City
	State
	Zip
	Country
	DaytimePhone
	EmailAddress
	GiftAmount
	DataComplete
	AccountNumber
	CreditCard
	ExpDateMonth
	ExpDateYear
	LastUpdateDate

The Prospect table contains the following fields:

Field Name	Data Type	Primary Key
Id	int	Yes
FirstName	varchar(50)	Foreign Key
LastName	varchar(50)	
MiddleInitial	varchar(50)	
Address	varchar(50)	
Apartment	varchar(50)	
City	varchar(50)	
State	varchar(50)	
Zip	varchar(50)	
Country	varchar(50)	
DaytimePhone	varchar(50)	
EmailAddress	varchar(255)	
GiftAmount	varchar(50)	
DataComplete	bit	
AccountNumber	varchar(50)	
CreditCard	varchar(50)	
ExpDateMonth	int	
ExpDateYear	int	
LastUpdateDate	smalldatetime	

The MZUser table contains the following fields:

Field Name	Data Type	Primary Key
Id	int	Yes
Username	varchar(50)	
RoleId	int	Foreign Key
FirstName	varchar(50)	
LastName	varchar(50)	
Email	varchar(255)	
Password	varchar(50)	

The MZRole table contains the following fields:

Field Name	Data Type	Primary Key
Id	int	Yes
Name	varchar(50)	
Description	varchar(255)	

The database contains the following stored procedures:

Procedure Name	Parameters	Return Value
sp_AddMZRole	@Name varchar(50), @Description varchar(255)	Id
sp_AddMZUser	@Username varchar(50), @FirstName varchar(50), @LastName varchar(50), @Email varchar(255), @Password varchar(50), @RoleId int	Id
sp_AddProspect	@FirstName varchar(50), @LastName varchar(50), @MiddleInitial varchar(50), @Address varchar(50), @Apartment varchar(50), @City varchar(50), @State varchar(50), @Zip varchar(50), @Country varchar(50), @DaytimePhone varchar(50), @EmailAddress varchar(255), @GiftAmount varchar(50), @DataComplete bit, @AccountNumber varchar(50), @CreditCard varchar(50), @ExpDateMonth int, @ExpDateYear int, @LastUpdateDate smalldatetime	Id
sp_UpdateMZRole	@Id int, @Name varchar(50), @Description varchar(255)	
sp_UpdateMZUser	@Id int, @Username varchar(50), @FirstName varchar(50), @LastName varchar(50), @Email varchar(255), @Password varchar(50), @RoleId int	

Procedure Name	Parameters	Return Value
sp_UpdateProspect	@Id int, @FirstName varchar(50), @LastName varchar(50), @MiddleInitial varchar(50), @Address varchar(50), @Apartment varchar(50), @City varchar(50), @State varchar(50), @Zip varchar(50), @Country varchar(50), @DaytimePhone varchar(50), @EmailAddress varchar(255), @GiftAmount varchar(50), @DataComplete bit, @AccountNumber varchar(50), @CreditCard varchar(50), @ExpDateMonth int, @ExpDateYear int, @LastUpdateDate smalldatetime	

13.5 DISCUSSION

The Mobile Zoo application provides an example of a Pocket PC application that utilizes web pages hosted on a Pocket PC using PWH. This type of application gives us the ease and flexibility associated with developing thin client web applications, along with the ability to store and forward data even when the mobile client is only intermittently connected to the back-end web server.

There are other advantages to developing a mobile application using web technology over Windows Forms. For example, many developers can create web pages more quickly and easily than they can develop Windows Forms applications. It is also probably easier to manage and maintain multiple sets of web pages than a web site with a separate fat client application.

This scenario can be applied to many situations in which a person has to gather information while disconnected from the server, but needs to synchronize that information with the server application later. Some examples are described below.

News and entertainment. This type of mechanism can be used to display and gather information from news, entertainment, and advertisement web pages on a Pocket PC that may be disconnected from the network.

Tourist walking tours. Web page hosting could also be used to provide Pocket PC versions of self-guided walking tours for major cities. For example, a walking tour of New York City could include web pages with pictures of landmark buildings and other attractions.

13.6 SUMMARY

This chapter described the development of a mobile client application that utilizes PWH to host web pages on a Pocket PC. We initially described the application's requirements using UML diagrams and flowcharts. We then described the architecture of the overall solution along with the detailed design of the client and server applications. Finally, we discussed using web page hosting on Pocket PCs.

APPENDIX A

Further Reading

In the sections below, we list a set of books and online resources that we found informative and useful, when designing and developing mobile applications. Please note that the comments describing each web site are our own, and do not represent the company in any official capacity. Interested readers should refer to the company's official web site for further details.

A.1 BOOKS

Albahari, Ben, Peter Drayton, and Brad Merrill. *C# Essentials*. O'Reilly, 2001.

Barwell, Fred et al. *Professional VB.NET*. WROX Press, 2001.

Boehm, Barry W. "A Spiral Model of Software Development and Enhancement." *IEEE Computer*, May 1988, pp. 61–72.

Booch, Grady, Ivar Jacobson, and James Rumbaugh. *The Unified Modeling Language User Guide*. Addison-Wesley, 1999.

Brans, Patrick. *Mobilize Your Enterprise—Achieving Competitive Advantage through Wireless Technology*. Prentice Hall PTR, 2003.

Forsberg, Christian and Andreas Sjöström. *Pocket PC Development in the Enterprise*. Addison-Wesley, 2001.

Grattan, Nick. *Pocket PC, Handheld PC Developer's Guide with Microsoft Embedded Visual Basic*. Prentice Hall PTR, 2001.

Gunnerson, Eric. *A Programmer's Introduction to C#*. APress, 2000.

Horstmann, Cay S. and Gary Cornell. *Core Java 2, Volume I-Fundamentals*. Sun Microsystems Press, 2001.

Jarrett, Rob and Philip Su. *Building Tablet PC Applications*. Microsoft Press, 2003.

Krell, Bruce E. *Pocket PC Developer's Guide*. McGraw-Hill Osborne Media, 2002.

Kruchten, Philippe. *The Rational Unified Process: An Introduction*, 2d ed. Addison-Wesley, 2000.

Martin, Didier et al. *Professional XML.* WROX Press, 2000.
Microsoft. *Analyzing Requirements and Defining Microsoft .NET Solution Architectures.* Microsoft Press, 2003.
Microsoft. *Building Secure Microsoft ASP.NET Applications.* Microsoft Press, 2003.
Oestereich, Bernd. *Developing Software with UML.* Addison-Wesley, 1999.
Richter, Jeffrey. *Applied Microsoft .NET Framework Programming.* Microsoft Press, 2002.
Smith, Ben and Brian Komar. *Microsoft Windows Security Resource Kit.* Microsoft Press, 2003.
Tacke, Chris et al. *Embedded Visual Basic: Windows CE and Pocket PC Mobile Applications.* SAMS, 2001.
Tiffany, Rob. *Pocket PC Database Development with Embedded Visual Basic.* APress, 2001.

A.2 ONLINE RESOURCES

http://www.avantgo.com

AvantGo, which is owned by iAnywhere, provides news and other information services from its web servers to mobile devices such as Pocket PCs.

http://www.bbc.co.uk

The British Broadcasting Corporation web site is a world-class resource for world and technology news, along with a variety of other services. Many of the BBC's web sites and services are also available on mobile devices. For example, you may synchronize BBC content between your Pocket PC and Desktop or Laptop PC using AvantGo.

http://www.ericsson.com

Ericsson manufactures and sells cellular telephones and Smartphones. Ericsson also supplies operators and service providers with end-to-end mobile and broadband Internet solutions.

http://www.fujitsu.com

Fujitsu is a hardware, services, software, and peripherals vendor. Products of interest include Pocket PCs, Tablet PCs, Laptop PCs, and a variety of enterprise-level servers. Other topics of interest include mobility and wireless, software and services, and platforms and electronic devices.

http://www.gartner.com

Gartner is a research and advisory firm that helps companies leverage technology to achieve business success. Gartner's businesses consist of research, consulting, measurement, events, and executive programs. The company's reports are highly regarded and utilized within many industries.

http://www.howstuffworks.com

How Stuff Works is a wonderful general reference source for researching types of technologies and how they work.

http://www.hp.com

Hewlett-Packard is a global provider of products, technologies, solutions, and services to consumers and businesses. Its offerings span information technology infrastructure, personal computing and access devices, global services, and imaging and printing. Products of interest include Pocket PCs (HP iPAQs), Tablet PCs, and Laptop PCs, a variety of enterprise level servers, and HP OpenView. Other topics of interest include mobility and wireless.

http://www.ianywhere.com

iAnywhere offers a line of mobile enterprise solutions designed to extend the reach of enterprise applications to mobile users. Its mobile enterprise portfolio ranges from infrastructure software to targeted business solutions and applications that address specific business needs, such as sales force automation, mobile email, and device and application management. iAnywhere is a subsidiary of Sybase.

http://www.ibm.com

IBM is a hardware, services, software, and peripherals vendor. Products of interest include Laptop PCs, a variety of enterprise-level servers, IBM WebSphere, and Rational Rose. Other topics of interest include mobility and wireless.

http://www.idc.com

IDC is a premier global market intelligence and advisory firm in the information technology and telecommunications industries. IDC analyzes and predicts technology trends so that their clients can make strategic, fact-based decisions on IT purchases and business strategy. The company's reports are highly regarded and utilized within many industries.

http://www.intel.com

Intel manufactures and supplies the computing and communications industries with chips, boards, systems, and software building blocks of computers, servers, and networking and communications products.

http://www.mercuryinteractive.com

Mercury Interactive is the vendor for a variety of products, including Load Runner for performance testing and SiteScope and SiteSeer for site monitoring.

http://www.microsoft.com

Microsoft is a software and services vendor. Products of interest include Microsoft Windows Mobile 2003, Microsoft Windows XP, Microsoft Windows XP Tablet Edition, and Microsoft Windows Server 2003.

http://www.microsoft.com/resources/practices

The Microsoft Patterns and Practices web site is a valuable resource for a number of topics, including architecture, application design, and security.

http://msdn.microsoft.com
We highly recommend the Microsoft Developer Network site for a large number of topics, including architecture, application development, .NET languages, mobile development, security, and the Microsoft Solutions Framework.

http://www.nokia.com
Nokia is a world leader in mobile communications. The company is a leading supplier of mobile phones as well as mobile, fixed broadband, and IP networks.

http://www.newburynetworks.com
Newbury Networks provides a wireless detection system that allows IT professionals to detect, monitor, and secure 802.11-based Wireless LANs by location.

http://www.oracle.com
Oracle is a software and services vendor. Products of interest include Oracle 9iAS and Oracle 9i database.

http://otn.oracle.com
The Oracle Technical Network site provides useful information for application development using Oracle, Sun Java, and J2EE-oriented products.

http://www.palm.com
Palm is a leader in mobile and wireless Internet solutions and is a leading provider of handheld computers.

http://www.rim.com
Research In Motion (RIM) is a leading designer, manufacturer, and marketer of wireless solutions for the mobile communications market. RIM provides platforms and solutions for access to information, including email, phone, SMS messaging, Internet, and intranet-based applications. RIM technology also enables a broad array of third-party developers and manufacturers to enhance their products and services with wireless connectivity to data.

http://rsasecurity.com
RSA Security provides solutions for establishing online identities, access rights, and privileges for people, applications, and devices.

http://www.sybase.com
In the past, Sybase was perhaps better known for its enterprise-class databases and software. Today, Sybase provides databases, software, services, and subsidiaries that help enterprises run mobile applications on a variety of mobile devices.

http://www.symantec.com
Symantec provides a range of content and network security software and appliance solutions to individuals, enterprises, and service providers. The company is a leading provider of security solutions for virus protection, firewall and virtual private network, vulnerability management, intrusion detection, Internet content and email filtering, and remote management technologies and security services to enterprises and service providers worldwide.

http://www.symbian.com

Symbian is a software licensing company that supplies the open, standard operating system (Symbian OS) for data-enabled mobile phones. Symbian is owned by a group of major manufacturers, including Ericsson, Nokia, Panasonic, Motorola, Psion, Samsung Electronics, Siemens, and Sony Ericsson.

http://www.symbol.com

Symbol Technologies manufactures a variety of products, including laser bar code scanners, handheld computers, and wireless communications networks for voice and data. The company also provides professional services, support, education, and training to its customers.

http://www.verizon.com

Verizon provides a wide variety of voice and data communications services that are both wired and wireless.

http://www.yahoo.com

Yahoo! is a great site for a variety of areas including world and technology news, email, games, groups, and a variety of other services. Many Yahoo! sites and services are also available on mobile devices. For example, you may synchronize Yahoo! content between your Pocket PC and Desktop or Laptop PC using AvantGo.

APPENDIX B

Pocket Web Host Design

The Pocket Web Host (PWH) is a "mini" web server that allows you to host web pages on a Pocket PC. In this section of the book, we will describe its architecture and the steps required to develop web pages for hosting on a Pocket PC. Readers who are interested in specific details may download the full code listing from the companion web site.

B.1 ARCHITECTURE

This section describes the architecture of PWH.

B.1.1 Hosting Web Pages

Consider a common scenario in which you are asked to develop an application that displays and gathers information to and from a mobile client. If the mobile client is always connected to the server, a thin client is typically used and web pages are hosted on the web server for display on the mobile device. If the mobile client is partially connected to the web server, a fat client application that uses Microsoft Windows Forms may be developed instead.

Developing a fat client using Microsoft Windows Forms, however, may not be a desirable situation for several reasons. For example, if your existing server application is already web-based, you may not want to develop a set of forms that use a different technology. Furthermore, if you have two or more different mobile device types to support, such as a Pocket PC and a Tablet PC, you would have to develop separate sets of code for each device type as well as maintain a third set of code for the server.

However, if you were able to host web pages on mobile devices, such as Pocket PCs, it would be possible to develop separate sets of web pages, instead of Microsoft Windows Forms. This would be much easier to develop and maintain. In addition, web pages are much more portable between device types (e.g., Pocket PC, Tablet PC) than Windows Forms.

For reasons such as these, it may be very beneficial to be able to host web pages on multiple device types. In essence, we are looking for a mini web server that can execute web pages and provide some of the ASP.NET functionality on the Pocket PC. Thus, we are looking for something equivalent to Microsoft IIS and Microsoft ASP.NET on a Pocket PC.

Unfortunately, there seem to be very few commercially available products that are able to host web pages on a Pocket PC in the fashion we describe. For example, Microsoft has released an HTTP Server that can process simple ASP web pages on a Pocket PC but it does not currently support ASP.NET. This poses a problem since Microsoft is discouraging application development for Microsoft Windows Mobile 2003 that is not based on .NET. For example, Microsoft encourages all data access to be done with ADO.NET and will no longer support the use of ADOCE.

As a result, we developed our own mini web server program, called the Pocket Web Host (PWHOST), and an associated program, called the Pocket Web Host Complier (PWHCOMP). Essentially, the PWH environment re-creates and re-leverages some of the Microsoft ASP.NET compile-time and run-time technologies on a Pocket PC.

B.1.2 Environment

Normally, ASP.NET web pages on a Microsoft Windows server are serviced through the use of Microsoft IIS and ASPNET_WP.EXE (see Figure B–1). Microsoft IIS operates as the web server while the ASPNET_WP.EXE process serves as both a compiler and type loader of ASP.NET pages and code.

In order to emulate these activities, the PWH environment was developed with PWHOST and PWHCOMP jointly providing this functionality (see Figure B–2). In this flow, PWHOST acts as a web server and type loader on the Pocket PC and is able to run and service web pages on a Pocket PC with the .PWH extension. The web pages are then viewable through a standard Pocket PC web browser. PWHCOMP, in contrast, is used by developers on the server to compile PWH pages.

B.1.3 PWHCOMP

Each PWH page requires two code-behind files in order to be hosted by the PWHOST environment. The first is the standard, user-created code-behind, which contains the OnInit and PageLoad events. The second code-behind is automatically generated by PWHCOMP, which carries out the following operations:

1. Locate and process the <PWH:> tags in the given web page. The locations of these identifiers are then placed in a linked list data structure.

2. Construct a generated code-behind. This code-behind will call the user-created code-behind's OnInit and PageLoad events. It writes out the HTML portion of the web page and renders the supported controls. At present, HTML is completely supported, but only one custom HTML control is supported.
3. Generate an assembly manifest.

The result is a C# class file that is responsible for rendering the web page and invoking the OnInit and PageLoad events of the user-created code behind.

Figure B–1 Microsoft IIS and Microsoft ASP.NET relationship

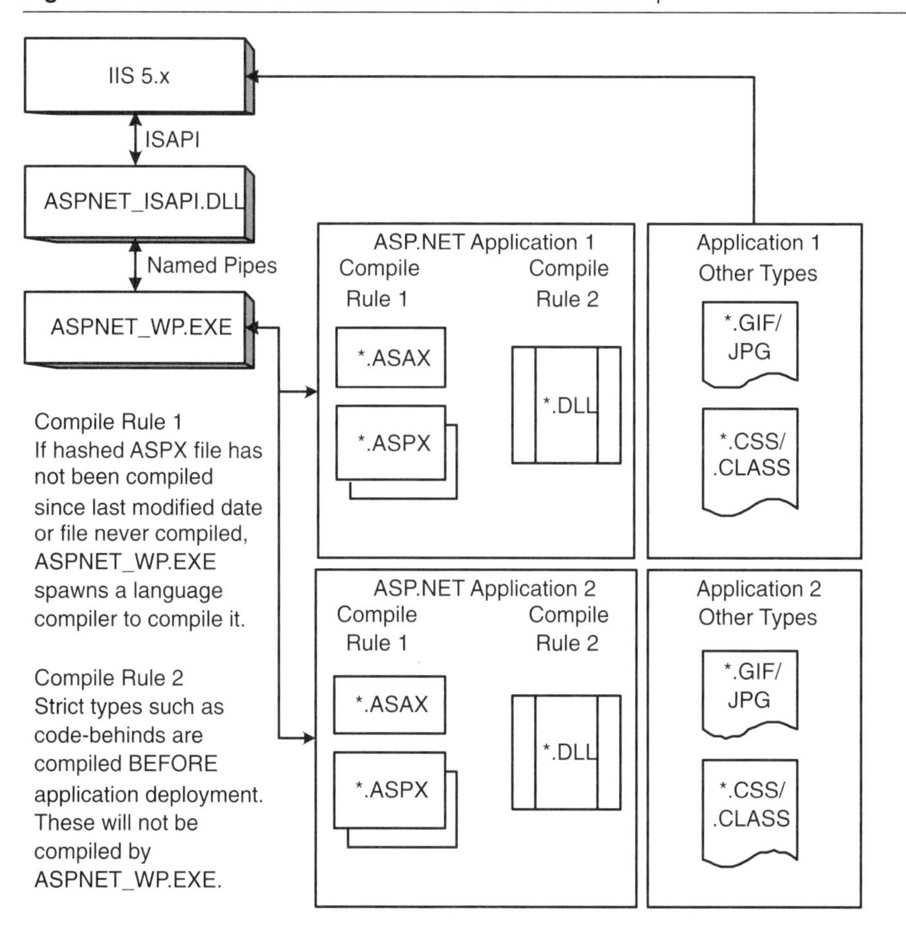

Figure B–2 PWHOST environment execution and flow paths

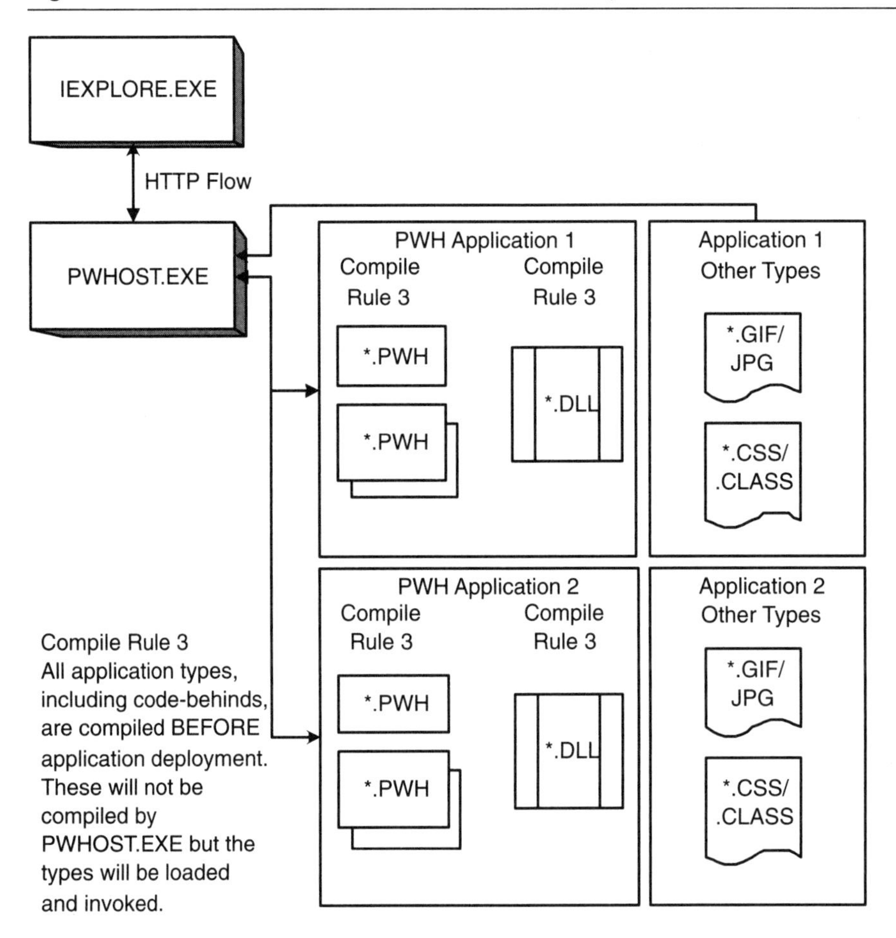

B.1.4 PWHOST

The process pipeline flow for a PWH page with a code-behind is as follows:

1. Determine which web page has been requested and load that page's generated code-behind.
2. The generated code-behind calls the user-created code-behind's OnInit and PageLoad methods, passing an array list of environment variables. This array list contains the HTTP Request object and all the controls from the page.
3. The user-created code-behind may set the values of the controls in its OnInit and PageLoad events.
4. After processing the user-created code-behind page, the generated code-behind continues processing the PWH stream. It writes out the HTML portion of the PWH page and renders the supported controls.
5. The generated code-behind then returns the buffered results to PWHOST, which renders the page as HTML and is viewable in the Pocket PC web browser.

B.2 DEVELOPMENT PROCEDURE

In the sections below, we describe the procedure for developing web pages that can run in the PWH environment. We start with the general procedure, followed by a simple example.

B.2.1 General Procedure

Developers will typically write a set of PWH pages, pass them through PWHCOMP, test, release, and run them using PWHOST (see Figure B–3).

PWHOST allows web pages to be hosted and serviced on Pocket PCs even when the Pocket PCs are disconnected from a host web server. Information gathered through web pages on the Pocket PC is stored in a local database. When a connection to a server is re-established, the mobile application may synchronize the data with the server. Note that the synchronization is not done by PWHOST. Each web application is responsible for synchronizing its own data. Typically, we use a web service for Pocket PC clients that need to synchronize their data with a server.

Figure B–3 PWH general development procedure

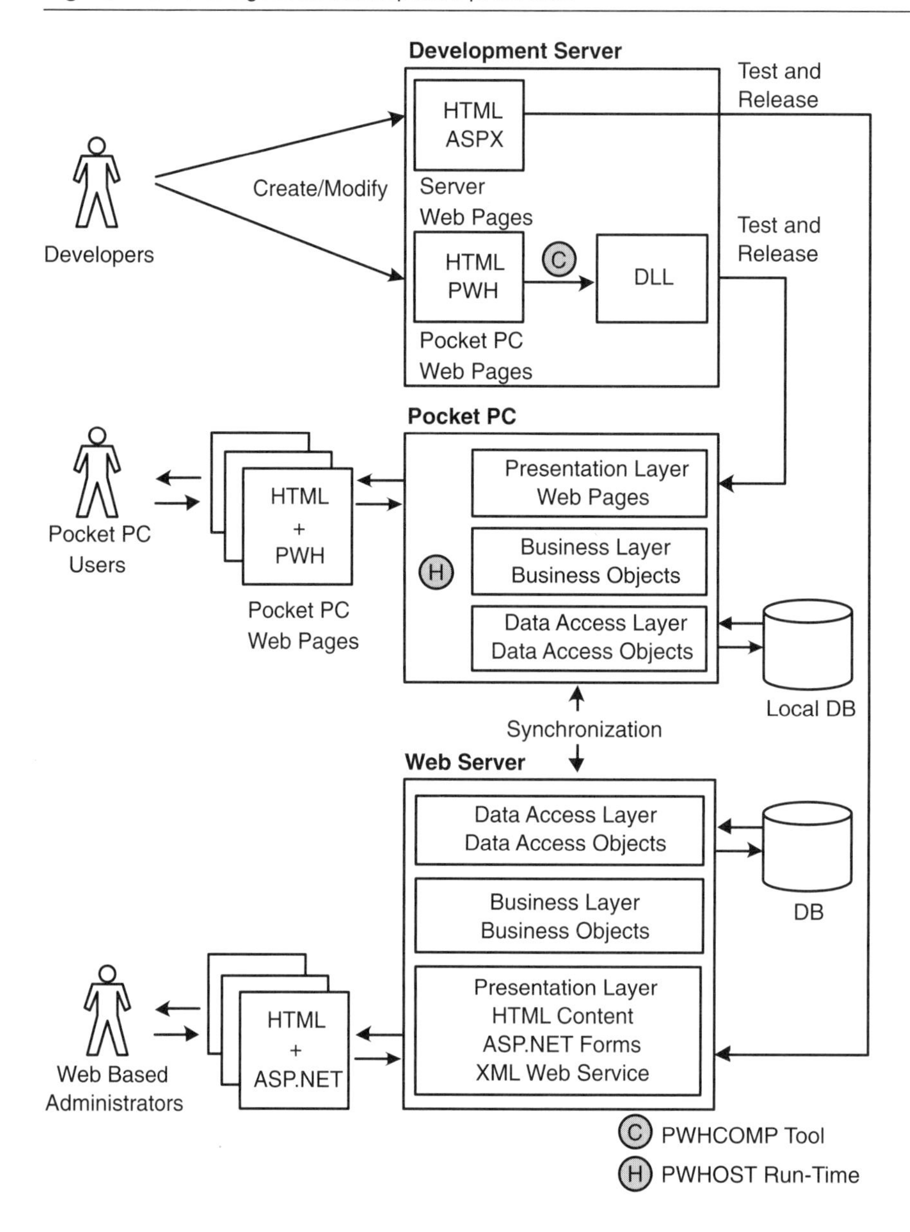

B.2.2 Example

In the following sections, we will use a simple program ("HelloWorld") to describe the PWH development procedure in more detail (see Figure B–4). The program files are located in the HelloWorld directory under the PWH folder in the source code accompanying this book.

1. **Write the PWH page and code-behind.** Use a text editor or Microsoft Visual Studio.NET 2003 to write the following page and its associated code-behind:

   HelloWorld.pwh. At this time, the only supported control is our custom HTML Control. This is similar in functionality to the ASP.NET label control but is a separate entity and should not be confused with it. Our custom HTML control must be specified as follows:

```
<PWH:HTMLCTRL id="lblMessage" runat="server"></PWH>
```

Figure B–4 PWH development flow

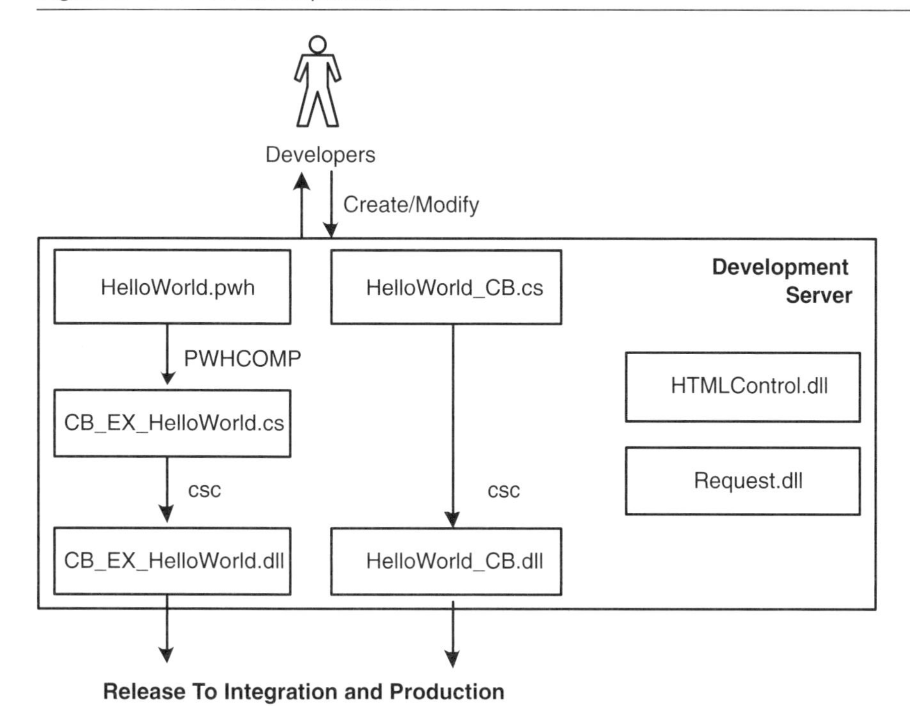

The contents of the HelloWorld.pwh file are as follows:

```
<HTML>
  <HEAD>
          <title>Hello World</title>
   </HEAD>
   <body>
          <form id="Form1" method="post" runat="server">
                <TABLE id="maintable" cellSpacing="0"
        cellPadding="0" width="600" border="0">
                       <TR>
                              <TD align="left">
                                     <h1>Hello World</h1>
                                     <br>
                                     First Name: <input
        type=text name=firstname value=''>
                                     <br>
                                     Last Name: <input
        type=text name=lastname value=''>
                                     <br>
                                     <input type=Submit
        value=Submit>
                                     <br><br>
                                     <PWH:HTMLCTRL
        id="lblMessage" runat="server"></PWH>
                               </TD>
                       </TR>
                </TABLE>
          </form>
  </body>
</HTML>
```

HelloWorld_CB.cs. The contents of HelloWorld_CB.cs (the HelloWorld code-behind) are as follows:

```
/*
'*************************************************************
          *******************
' FileName: HelloWorld_CB.cs
' Author: VL, HS, RS
' Created: Nov-2003
' The code behind file for HelloWorld.pwh
'*************************************************************
          ******************
*/
using System;
using System.Collections;
using Controls;
using PwhSystemInput;
namespace HelloWorld
{
  /// <summary>
  /// Summary description for codebehind.
  /// </summary>
  public class HelloWorld_CB
  {
         protected HtmlControl lblMessage;
         protected HTTPRequest Request;
         public HelloWorld_CB()
         {
                //
                 // TODO: Add constructor logic here
                 //
          }
          public void OnInit(ArrayList CtlContainer)
          {
                 Request = (HTTPRequest)CtlContainer[0];
                 lblMessage = (HtmlControl)CtlContainer[1];

}


          public void Page_Load()
```

```
            {
                    String firstname =
            Request.Form("firstname");
                    String lastname =
            Request.Form("lastname");
                    if(firstname==null)
                    {
                            lblMessage.ControlText="Please enter
            your first and last name";

    }
                    else
                    {
                            lblMessage.ControlText = "Hello " +
            firstname + " " + lastname + "!!!!";
                    }
            }
      }
    }
```

2. **Generate hidden code behind using PWHCOMP.** PWHCOMP.EXE generates a hidden code-behind for each PWH file. Copy your HelloWorld.pwh file from the HelloWorld directory to the bin\debug directory of the PWHComp project. Open a command prompt window and enter the following:

```
> pwhcomp.exe HelloWorld.pwh HelloWorld HelloWorld_CB
```

"HelloWorld.pwh" is the source file name while "HelloWorld" refers to the code-behind namespace and "HelloWorld_CB" refers to the code-behind class name. This command generates the output C# file CB_EX_HelloWorld.cs, whose contents are as follows:

```
using System;
using System.Text;
using System.Collections;
using System.Collections.Specialized;
using Controls;
using PwhSystemInput;
using System.Reflection;
```

```
using HelloWorld;
#region AutoGenerated Code
//using users_CB;
#endregion
namespace CB_EX_HelloWorld
{
  public class res_buffer
  {
         String s_all = "";

         ArrayList ctlContainer=new ArrayList();

         public void Write(string s)
         {
                 s_all=s_all + s;
         }
         public string GetStream()
         {
                 return s_all;
         }
         public string ctlVal(string s_in, string s_ctlID)
         {
                 int i_pos;
                 i_pos = s_in.IndexOf(s_ctlID,0);
                 if (i_pos != -1)
                 {
                        return
         s_in.Substring(i_pos,s_ctlID.Length+1);
                 }
                 else
                 {
                        return null;
                 }
          }
          public void Advise(string s_ctl, object o)
          {
                 ctlContainer.Add( o );
          }
          public string UnAdvise(object o)
          {
```

```
                return null;
            }
            public ArrayList GetCtlContainer()
            {
                return ctlContainer;
            }
    }
            // This is the generated code-behind
    public class HTTPPage
    {
            res_buffer Response;
            HTTPRequest Request= new HTTPRequest();
            public string Init(string sin, string sout)
            {
                // Init stuff we need right here.
                Response = new res_buffer();

                Request.SetupRequest(sin);
                InvokeRender();
                return Response.GetStream();


            }
            public void InvokeRender()
            {
                //for each entity in the array, write it out
          to the buffer
                //or call invoke on the code-behind
                #region AutoGenerated Code
                Response.Advise ("Request",Request);
                // AutoGEN Region
                //response.write string region.

HtmlControl lblMessage_lbl = new HtmlControl();
lblMessage_lbl.ControlText="";
Response.Advise("lblMessage_lbl",lblMessage_lbl);
string path =
        System.IO.Path.GetDirectoryName(System.Reflection.Assem
        bly.GetExecutingAssembly().GetName().CodeBase );
string assemblyname = path + "\\HelloWorld_CB.dll";
```

```
Assembly a = Assembly.LoadFrom(assemblyname);
Type page = a.GetType("HelloWorld.HelloWorld_CB");
object obj = Activator.CreateInstance(page);
MethodInfo oninit = page.GetMethod("OnInit");
MethodInfo pageLoad = page.GetMethod("Page_Load");
object[] paramArray = new object[1];
paramArray[0] = Response.GetCtlContainer();
object ret_obj = oninit.Invoke(obj, paramArray);
ret_obj = pageLoad.Invoke(obj, null);
Response.Write(@"<HTML>
  <HEAD>
          <title>Hello World</title>
  </HEAD>
  <body>
          <form id=""Form1"" method=""post"" runat=""server"">
                <TABLE id=""maintable"" cellSpacing=""0""
        cellPadding=""0"" width=""600"" border=""0"">
                       <TR>
                              <TD align=""left"">
                                     <h1>Hello World</h1>
                                     <br>
                                     First Name: <input
        type=text name=firstname value=''>
                                     <br>
                                     Last Name: <input
        type=text name=lastname value=''>
                                     <br>
                                     <input type=Submit
        value=Submit>
                                     <br><br>
                              ");
Response.Write (lblMessage_lbl.RenderControl());
Response.Write(@"                             </TD>
                       </TR>
                </TABLE>
          </form>
  </body>
</HTML>
");
```

```
                        //response.write invoke
                        //response.write
                        #endregion

            }
    }
    }
```

3. **Compile using C# compiler.** Once you have generated the hidden code-behind, compile both your user-created code-behind and the generated code-behind using the normal C# compiler. This is done by opening a Visual Studio command prompt by going to Start Menu > Programs > Microsoft Visual Studio .NET 2003 > Visual Studio .NET Tools > Visual Studio .NET 2003 Command Prompt. Then navigate to the HelloWorld directory and type the following:

```
> csc /reference:HTMLControl.dll /reference:Request.dll
        /target:library HelloWorld_CB.cs

> csc /reference:HTMLControl.dll /reference:Request.dll
        /reference:HelloWorld_CB.dll /target:library
        CB_EX_HelloWorld.cs
```

   HTMLControl.dll and Request.dll are external dependent files that we have provided.
4. **Deploy.** Initially, you must deploy the PWHOST program. This can be done from within Visual Studio .NET. PWHOST is deployed into \Program Files\pwhost on the mobile device. You must then copy the entire HelloWorld folder into \Program Files\pwhost on the device so that all the files are in \Program Files\pwhost\HelloWorld. If you are using an emulator, you must share the directory on your development machine and then map the share from the emulator to copy the files over.
5. **Run.** Open Internet Explorer on the Pocket PC. PWHOST runs on port 8080, so enter "http://127.0.0.1:8080/HelloWorld/index.html" as the URL. (If you are using Pocket PC 2002, you may enter "localhost" as the server name instead of the local IP address, but this does not work properly on Microsoft Windows Mobile 2003. If you are using Microsoft Windows Mobile 2003 you should enter "127.0.0.1" instead.)

   Entry of this URL will bring up an HTML page with a link to HelloWorld.pwh (see Figure B–5). Click that link and you will see a screen with "Hello World" and two text boxes. Enter your first and last name. Upon form submission, PWHOST will service the page and return and display the full name back on the Web Form (see Figure B–6).

Figure B–5 Hello World HTML page

Figure B–6 Hello World PWH Page

B.3 SUMMARY

PWH is a "mini" web server that allows you to host web pages on a Pocket PC. This mechanism allows you to service web pages on a Pocket PC even when it is disconnected from a network.

There are a number of similar ideas and technologies on the market worth noting. AvantGo, for example, allows Pocket PC users to download static HTML pages and perform HTML form submissions while disconnected. Any submitted information is subsequently uploaded when connectivity is re-established. This is a clever implementation but it also has limited ASP.NET processing capability at present.

It would perhaps be ideal if a product similar to PWH was implemented and supported by a software vendor. Until that time, you may find it useful to treat PWHOST as a starting point and extend its capabilities so that it provides full ASP.NET and perhaps even CGI, JSP, and Java support.

INDEX

H

I

K

L

M

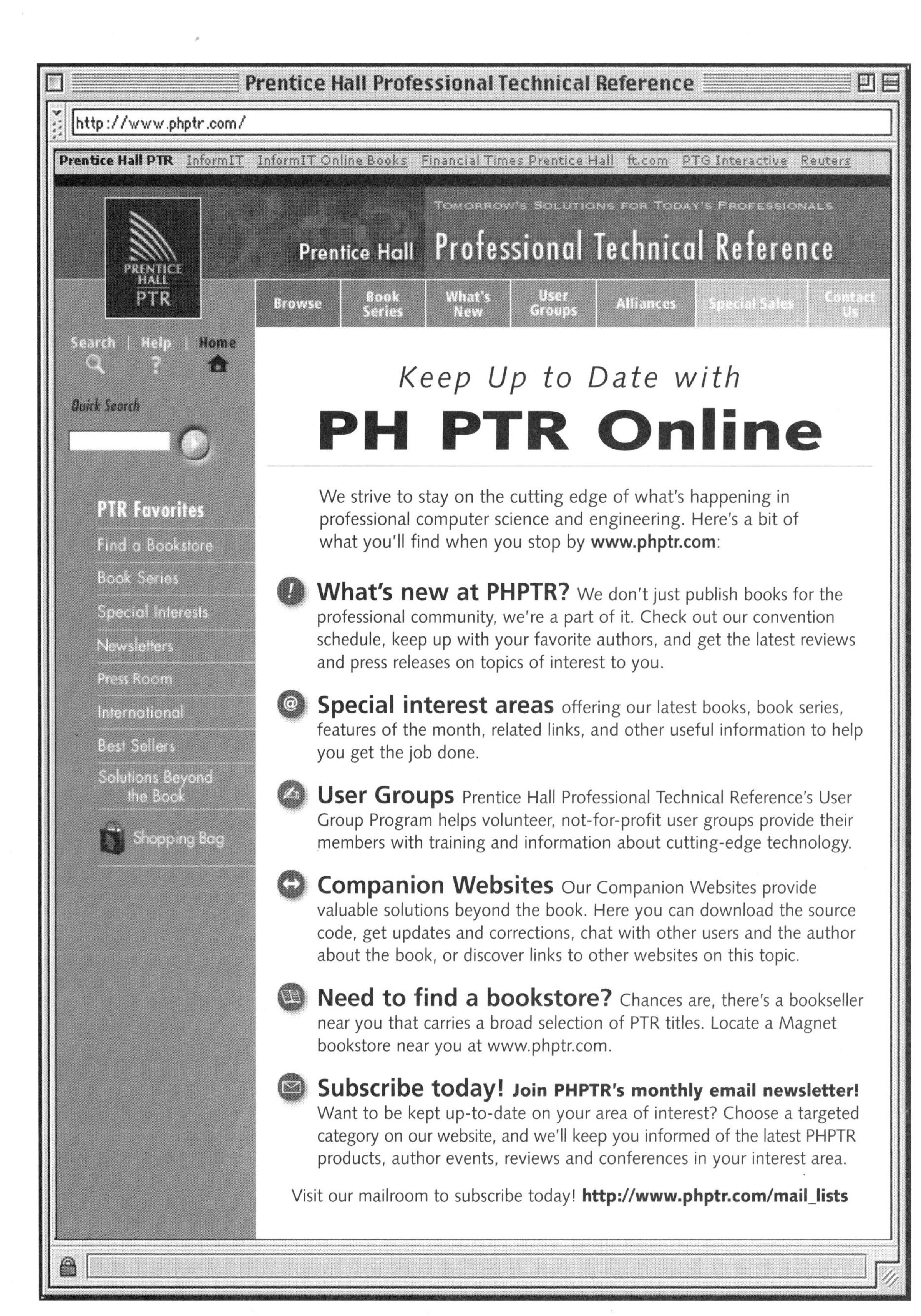
Prentice Hall Professional Technical Reference
http://www.phptr.com/
Prentice Hall PTR
InformIT
InformIT Online Books
Financial Times Prentice Hall
ft.com
PTG Interactive
Reuters
TOMORROW'S SOLUTIONS FOR TODAY'S PROFESSIONALS
PRENTICE HALL PTR
Prentice Hall Professional Technical Reference
Browse
Book Series
What's New
User Groups
Alliances
Special Sales
Contact Us
Search | Help | Home
Quick Search
PTR Favorites
Find a Bookstore
Book Series
Special Interests
Newsletters
Press Room
International
Best Sellers
Solutions Beyond the Book
Shopping Bag
Keep Up to Date with
PH PTR Online
We strive to stay on the cutting edge of what's happening in professional computer science and engineering. Here's a bit of what you'll find when you stop by www.phptr.com:
What's new at PHPTR? We don't just publish books for the professional community, we're a part of it. Check out our convention schedule, keep up with your favorite authors, and get the latest reviews and press releases on topics of interest to you.
Special interest areas offering our latest books, book series, features of the month, related links, and other useful information to help you get the job done.
User Groups Prentice Hall Professional Technical Reference's User Group Program helps volunteer, not-for-profit user groups provide their members with training and information about cutting-edge technology.
Companion Websites Our Companion Websites provide valuable solutions beyond the book. Here you can download the source code, get updates and corrections, chat with other users and the author about the book, or discover links to other websites on this topic.
Need to find a bookstore? Chances are, there's a bookseller near you that carries a broad selection of PTR titles. Locate a Magnet bookstore near you at www.phptr.com.
Subscribe today! Join PHPTR's monthly email newsletter!
Want to be kept up-to-date on your area of interest? Choose a targeted category on our website, and we'll keep you informed of the latest PHPTR products, author events, reviews and conferences in your interest area.
Visit our mailroom to subscribe today! http://www.phptr.com/mail_lists